# THAILAND AS A DIGITAL NOMAD

## Turn Your Dream into Reality with This Step-by-Step Guide



By

Candice Rafferty

ISBN: 978-1-7637223-0-9 (e-book)
ISBN: 978-1-7637223-1-6 (paperback)



Front cover image by Lesla T.
Map design by Janik S.

Book design by Arslan.
Printed by Cetwixt Ink in Australia.
First printing edition 2024.

**Cetwixt Ink**
67 Jindabyne Road
Kingston Beach, Hobart
Tasmania, Australia, 7050

cetwixt.ink@gmail.com
@cetwixt.ink (Instagam)

# Table of Contents

# Reader Advisory

Before you commence reading this guide, please take note of the following important information. I thank you for your understanding regarding these limitations, and I hope you find this book a valuable resource.

- The content of this book is based on research conducted in 2024 and reflects my own personal knowledge and experiences. While every effort has been made to ensure the accuracy of the information provided, it should be read as an indicative guide only. Some of the information provided is subject to change or may have changed since the time of writing.
- As the author, I do not represent any company or entity, and I have not received any compensation for writing this book. All opinions expressed are solely my own. This book is intended to provide general information and insights; however, it should not be relied upon as a sole source of information for making any major decisions – whether they be financial, legal, medical, or otherwise.
- It is strongly recommended that you seek professional advice from qualified experts in the fields of medicine, law, accounting, or any other pertinent areas before taking any significant decisions or actions based on the data or opinions presented in this book. Sound professional guidance will help you make informed decisions that best suit your individual circumstances.
- In particular, prices listed for goods and services, whilst accurate as of November 2024, will be subject to change due to various factors, including but not limited to seasonal shifts, economic conditions, and market dynamics. Always verify current costs directly from

reliable and updated sources before making any significant financial or life decisions.

- By reading this book, you acknowledge and agree that the author and publisher cannot be held responsible for any errors, omissions, or changes in the information provided, nor for any decisions you make based on this content.

# About This Book

I have lived and worked in Asia for over 20 years, more than half that time in Thailand. I have personally gone through the experience and process of researching and living the contents of this book and now want to share my knowledge with others. This book is a heartfelt tribute to a country that has enriched my life in countless ways, and I am eager to share its wonders with you.

This book has been written as a handy go-to guide for all things Thailand. It does not need to be read chronologically; each chapter stands alone. So pick those of interest and dive right in!

For more details on—the places, services, and products mentioned within, handy links are provided in the **Resources** section at the back of the book.

Should you have any questions or comments about anything you've read or you'd like to collaborate on a project, please visit **@cetwixt.ink** on Instagram or email **cetwixt.ink@gmail.com**. Happy reading!

# Map

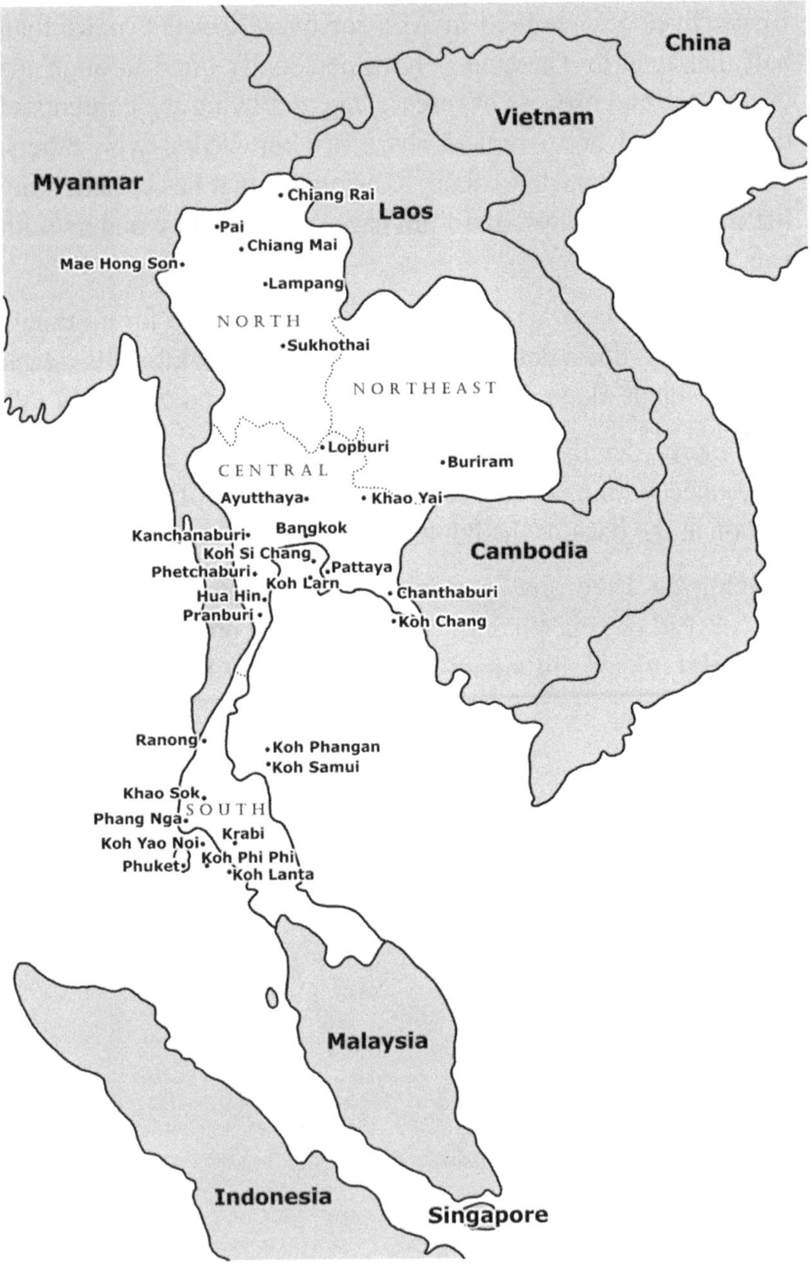
China
Vietnam
Myanmar
Chiang Rai
Laos
Pai
Chiang Mai
Mae Hong Son
Lampang
NORTH
Sukhothai
NORTHEAST
Lopburi
Buriram
CENTRAL
Ayutthaya
Khao Yai
Kanchanaburi
Bangkok
Koh Si Chang
Cambodia
Phetchaburi
Pattaya
Koh Larn
Chanthaburi
Hua Hin
Pranburi
Koh Chang
Ranong
Koh Phangan
Koh Samui
Khao Sok
SOUTH
Phang Nga
Krabi
Koh Yao Noi
Koh Phi Phi
Phuket
Koh Lanta
Malaysia
Indonesia
Singapore

# Introduction

Pack your bags, digital wanderer! You're on the brink of an adventure that will take you from the bustling skyscrapers of Bangkok to the serene beaches of Phuket, with a few spicy street food detours in between. This isn't just a guidebook; it's your survival kit for navigating the wonderfully chaotic, eternally enchanting world of Thailand as a digital nomad. And believe me, you'll need it.

Let's set the scene. You've just landed in Thailand. The air is thick with the scent of lemongrass and car exhaust, motorbikes whiz past like chaotic fish in a pond, and suddenly, you're hit with the realisation – this isn't Kansas anymore. But you're not just visiting; you're settling in, setting up shop, and maybe even calling this place home for a while. So, where do you begin?

**The Paperwork Dance**

First off, let's talk about the elephant in the room – Thai bureaucracy. Thailand loves its paperwork, and if you're not careful, you might find yourself tangled in red tape instead of tanning on the beach. Getting the right visa can feel like navigating a minefield; only instead of mines, you're dodging misunderstood forms and baffling regulations. But fear not! This book transforms the dreary details into a structured plan. You'll learn how to tackle the dreaded 90-day reporting process and actually understand the different visa options without needing a law degree.

**Home Sweet Bungalow**

Now, let's find you a place to lay your weary head. Whether it's a modern condo with a view of the Bangkok skyline or a cosy bungalow on Koh Samui, securing accommodation is an art form here. This book offers insider tips on everything from negotiating rent to understanding local quirks. And if communal living is your preference, we've got the information on the best co-living spaces that offer more than just a Wi-Fi connection – they offer a community.

**Wi-Fi and Workflows**

You might be sipping fresh coconut water, but you've got deadlines, meetings, and a boss who barely knows what time zone you're in. Thailand's co-working spaces are as diverse as its landscapes, and finding the right one can make all the difference in your work-life balance. This book will help you navigate these spaces – whether you need a private nook or a vibrant community hub – and ensure you always have a strong internet connection.

**Baht and Beyond**

Managing your finances in a new country can be akin to taming a wild beast. From setting up a bank account to understanding the cost of living, this book sheds light on how to access and spend your money wisely. Ever wondered why you keep getting strange looks when you try to tip like you do back home? We'll unravel that cultural mystery and many more. Plus, practical tips on avoiding scams and keeping your finances secure.

**Health is Wealth**

Hospital visits in a foreign country can be daunting. But in Thailand, healthcare ranges from surprisingly affordable to world-class. This book will guide you through the healthcare system, recommend the best insurance options, and even share tips on how to stay fit and healthy in the Land of Smiles. Consider this your ultimate wellness manual – because nobody wants a holiday to be a hospital stay. You might even pick up a new sport like Muay Thai or kitesurfing along the way.

**Cultural Riches**

You're not just here to work – you're here to experience a culture that's as rich and spicy as the local cuisine. This guide dives into the essential dos and don'ts, the fascinating festivals, and the charming quirks of Thai society. By the end, you'll not only blend in seamlessly but also appreciate the stunning diversity that makes Thailand truly special.

**Culinary Adventures**

You're not just here to work – you're here to experience a culture that's as rich and spicy as the local cuisine. This guide dives into the essential dos and don'ts, the fascinating festivals, and the charming quirks of Thai society. You'll discover the deep-rooted traditions that shape daily life and lend an extraordinary flavour to ordinary moments. By the end, you'll not only blend in seamlessly but also appreciate the stunning diversity that makes Thailand truly special.

**Tribe Building**

For all you social butterflies and introverted networkers, finding your tribe is crucial. Whether you're a solo explorer or relocating with family, chapters on community and social life will help you build meaningful connections. From joining local clubs to attending digital nomad meetups and community beach cleanups, we've got you covered for making friends who might just become lifelong companions.

**Little Wanderers**

Got kids? They're in for a treat. With a host of international schools, extracurricular activities, and homeschooling options, this book makes sure that your littlest nomads get the best education and social experiences Thailand has to offer. From cultural integration tips to creating a supportive home environment, you'll learn how to transition smoothly into local schools and ensure your kids are as happy and engaged as you are. Plus, discover hidden gems for family-friendly activities and weekend adventures.

**The Ultimate Playbook**

So, why is this book essential? Every tip, every suggestion, and every piece of advice is tailored to make your Thai adventure smoother, richer, and infinitely more enjoyable. It's not just about surviving; it’s about thriving. Whether you're here for a month, a year, or forever, consider this your ultimate playbook – a combination of sage advice and practical knowledge to make your journey successful.

So, ready to dive in? Your Thai adventure awaits, and with this book in hand, you're not just stepping into a new country – you're jumping headfirst into a spectacular new chapter of your life. Let the journey begin!

# Chapter 1: Living & Housing

## Finding Your Perfect Home in Thailand

So, you've decided to take the plunge and live the digital nomad life in Thailand – fantastic choice! One of the most exciting parts of your journey will be finding the perfect place to call home. From bustling city apartments to serene beachside bungalows, Thailand offers a diverse range of accommodation options to suit every taste and budget. In this chapter, we'll explore how to find short-term and long-term rentals, highlight popular neighbourhoods, break down the cost of living and introduce you to co-living and co-working spaces.

## HOME AWAY FROM HOME

Thailand has a plethora of amazing cities, towns, and islands in which to settle down and make your new home. Each has its own unique vibe and attractions, catering to different nomads depending on their lifestyle, budget, and preferences. Whether you're drawn to the bustling energy of Bangkok, the cultural richness of Chiang Mai, or the tranquil beaches of Koh Samui, there's a perfect spot for everyone. In this chapter, we'll explore the top ten places (alphabetically) to live as a digital nomad in Thailand, highlighting the key features that make each location special and ideal for remote work.

## Top Ten Places to Live in Thailand

### 1. Bangkok

Bangkok, the bustling capital of Thailand, is a vibrant metropolis that offers something for everyone. From its **bustling** street markets and cultural landmarks to its modern skyscrapers and trendy neighbourhoods, Bangkok is a city that never sleeps. It's a hotspot for digital nomads thanks to its fast internet, numerous co-working spaces, and lively expat community.

**Pros**

- Excellent infrastructure and amenities
- Wide variety of co-working spaces
- Thriving food scene and nightlife

**Cons**

- High cost of living compared to other Thai cities
- Traffic congestion and pollution
- The sheer volume of people, noise and heat can be overwhelming

**Neighbourhoods**

- **Ari**: A quieter, more residential area with a laid-back vibe. Ari is popular for its green spaces and local eateries. It's a great place for those who prefer a more relaxed lifestyle while still being close to the city centre.
- **Sathorn**: The business district, ideal for those who want to be close to co-working spaces and business hubs. Sathorn offers a mix of modern high-rises and traditional Thai houses, making it a diverse and dynamic area to live in.
- **Thonglor**: Known for its trendy cafes, restaurants, and nightlife, Thonglor is a favourite among young professionals and expats. It's a lively area with plenty of entertainment options and a cosmopolitan atmosphere.

**Suitable for**

- Affluent young professionals
- Families
- Urban enthusiasts

### 2. Chiang Mai

Chiang Mai, located in northern Thailand, is a favourite among digital nomads for its laid-back atmosphere, affordable cost of living, and vibrant expat community. The city is surrounded by mountains and lush countryside, offering plenty of opportunities for outdoor activities and exploration.

**Pros**

- Affordable cost of living
- Strong digital nomad community
- Beautiful natural surroundings

**Cons**

- Limited nightlife compared to Bangkok
- Seasonal air pollution (burning season from February to April)
- Can feel isolated from major business hubs

**Neighbourhoods**

- **Nimmanhaemin**: The heart of the digital nomad scene in Chiang Mai, Nimmanhaemin is packed with cafes, co-working spaces, and a vibrant expat community. It's a trendy area with a youthful vibe and plenty of modern conveniences.
- **Old City**: Offers a more traditional Thai experience with its temples, markets, and historic sites. The Old City is still close to many modern conveniences and offers a unique blend of old and new.
- **Santitham**: A quieter, more local area that's still close to Nimmanhaemin and the Old City. Santitham offers a more authentic Thai living experience with local markets, street food, and a slower pace of life.

**Suitable for**

- Budget-conscious professionals
- Families
- Nature enthusiasts

### 3. Chiang Rai

Chiang Rai, located in northern Thailand, is a charming city known for its beautiful temples, scenic landscapes, and cool climate. It's a great destination for digital nomads seeking a quieter and more culturally rich experience. The city offers a lower cost of living and a relaxed atmosphere.

**Pros**

- Affordable cost of living
- Beautiful temples and cultural attractions
- Easy access to Golden Triangle exploration

**Cons**

- Limited nightlife and entertainment options
- Less international cuisine options compared to larger cities
- Can feel isolated from the rest of the country.

**Neighbourhoods**

- **Ban Du**: A more suburban area located just outside the city centre, Ban Du offers a quieter and more local experience. It's a great place for those who prefer a slower pace and a more authentic Thai lifestyle.
- **City Centre**: The heart of Chiang Rai, the City Centre offers a mix of local markets, temples, and cafes. It's a convenient area to live in, with easy access to amenities and attractions.
- **Mae Fah Luang**: Located near Mae Fah Luang University, about 20km north of the city centre, this area

offers a mix of student life and local amenities. It's a vibrant and youthful neighbourhood with plenty of cafes and affordable accommodation options.

**Suitable for**

- Budget-conscious professionals
- Nature and adventure lovers
- Culture enthusiasts

**4. Hua Hin**

Hua Hin, located on the Gulf of Thailand, is a charming beach town that has become increasingly popular among digital nomads. Known for its relaxed atmosphere, beautiful beaches, and royal connections, Hua Hin offers a more laid-back alternative to the bustling cities of Bangkok and Pattaya.

**Pros**

- Relaxed and family-friendly atmosphere
- Beautiful beaches and outdoor activities
- Proximity to Bangkok (easy access for weekend trips)

**Cons**

- Limited nightlife and schooling options compared to larger cities
- Can be quieter, which may not suit those looking for a bustling social scene
- Fewer co-working spaces compared to major cities

**Neighbourhoods**

- **Cha-Am:** Located just north of Hua Hin, Cha-Am is a quieter beach town with a more local feel. It's a great place for those who prefer a slower pace and a more authentic Thai living experience.
- **Hua Hin Town:** The central area of Hua Hin offers a mix of local and expat amenities, including markets,

restaurants, and cafes. It's a convenient and lively area with easy access to the beach and other attractions.

- **Khao Takiab:** A scenic area located just south of Hua Hin town, Khao Takiab offers beautiful beaches, seafood restaurants, and a relaxed atmosphere. It's a great place for families and those who enjoy outdoor activities.

**Suitable for**

- Budget-conscious professionals
- Families
- Those seeking tranquillity

**5. Koh Chang**

Koh Chang, also known as Elephant Island, is the second-largest island in Thailand and is known for its beautiful beaches, lush jungles, and relaxed atmosphere. It's a popular destination for digital nomads seeking a peaceful and nature-focused lifestyle. The island offers a strong sense of community and a slower pace of life.

**Pros**

- Beautiful beaches and natural surroundings
- Peaceful and relaxed atmosphere
- Strong sense of community among expats

**Cons**

- Limited nightlife, entertainment, and schooling options
- Smaller expat community compared to larger islands
- Restricted public transport and healthcare options

**Neighbourhoods**

- **Klong Prao Beach**: A quieter and more laid-back area, Klong Prao Beach offers beautiful beaches and a relaxed atmosphere. It's a great place for families and those who prefer a slower pace.

- **Lonely Beach**: Known for its backpacker and hippie vibe, Lonely Beach offers a more budget-friendly option with a range of bars, hostels, and beachfront bungalows. It's a great place for younger travellers and those looking for a more bohemian experience.
- **White Sand Beach**: The most popular area on Koh Chang, White Sand Beach offers a range of amenities, including restaurants, shops, and accommodation options. It's a great place for those who want to be close to the action while enjoying beautiful natural surroundings.

**Suitable for**

- Budget-conscious professionals
- Beach lovers
- Those seeking a peaceful retreat

### 6. Koh Lanta

Koh Lanta is a serene island located in the Andaman Sea, known for its long sandy beaches, lush mangroves, and relaxed atmosphere. It's a popular destination for digital nomads seeking a peaceful and nature-focused lifestyle. The island offers a slower pace of life and a strong sense of community.

**Pros**

- Peaceful and relaxed atmosphere
- Beautiful beaches and natural surroundings
- Strong sense of community among expats

**Cons**

- Limited nightlife, entertainment, and schooling options
- Smaller expat community compared to larger islands
- Restricted healthcare facilities

**Neighbourhoods**

- **Klong Nin**: A quieter and more laid-back area, Klong Nin offers beautiful beaches and a relaxed atmosphere. It's a great place for those who prefer a slower pace and a more local experience.
- **Long Beach**: Known for its long stretch of sandy beach, Long Beach is a popular area for expats and digital nomads. It offers a mix of beach bars, restaurants, and accommodation options.
- **Saladan**: The main town on Koh Lanta, Saladan, offers a range of amenities, including restaurants, shops, and accommodation options. It's a convenient base for exploring the island.

**Suitable for**

- Budget-conscious professionals
- Beach lovers
- Those seeking a laid-back lifestyle

## 7. Koh Samui

Koh Samui is one of Thailand's largest islands, known for its palm-fringed beaches, luxury resorts, and vibrant nightlife. It's a popular destination for tourists and digital nomads, offering a balanced mix of relaxation and excitement. The island has a well-developed infrastructure and plenty of amenities to cater to expats.

**Pros**

- Beautiful beaches and luxury amenities
- Vibrant nightlife and social scene
- Wide range of accommodation options

**Cons**

- Higher cost of living compared to mainland Thailand

- Can be crowded and touristy
- Costly to get in and out by flying (limited accessibility due to airport monopoly)

**Neighbourhoods**

- **Bophut**: Known for its charming Fisherman's Village, Bophut offers a more laid-back atmosphere with boutique shops, beachfront restaurants, and a weekly night market. It's a popular area for families and those who prefer a quieter lifestyle.
- **Chaweng**: The most popular and lively area on Koh Samui, Chaweng offers a bustling nightlife, numerous restaurants, and plenty of shopping options. It's a great place for those who enjoy a vibrant social scene.
- **Lamai**: A quieter alternative to Chaweng, Lamai still offers plenty of amenities, including restaurants, bars, and shops. It's a good balance between excitement and relaxation.

**Suitable for**

- Budget-conscious professionals
- Beach lovers
- Resort enthusiasts

**8. Krabi**

Krabi, located on the Andaman coast, is famous for its stunning limestone cliffs, azure waters, and beautiful islands. It's a haven for outdoor enthusiasts and beach lovers. Krabi offers a more relaxed atmosphere than Phuket, making it a great choice for digital nomads seeking a quieter, nature-focused lifestyle.

**Pros**

- Stunning natural landscapes and outdoor activities
- Quieter and more relaxed than Phuket
- Affordable cost of living

**Cons**

- Limited nightlife, entertainment, and schooling options
- Smaller expat community compared to larger cities
- Limited public transport options

**Neighbourhoods**

- **Ao Nang**: The most popular area in Krabi, Ao Nang offers a mix of beach life, restaurants, and shops. It's a great place for those who want to be close to the action while still enjoying beautiful natural surroundings.
- **Krabi Town**: The provincial capital, Krabi Town, offers a more local experience with its markets, riverside restaurants, and laid-back vibe. It's a good base for exploring the region and offers affordable accommodation options.
- **Railay Beach**: Accessible only by boat, Railay Beach is a paradise for rock climbers and beach enthusiasts. It's a more remote and tranquil area, perfect for those who want to escape the hustle and bustle and be close to nature.

**Suitable for**

- Budget-conscious professionals
- Nature lovers
- Adventure seekers

**9. Pattaya**

Pattaya, located on the eastern Gulf coast of Thailand, is known for its vibrant nightlife, beautiful beaches, and lively expat community. It's a popular destination for digital nomads looking for a mix of work and play. Pattaya has a well-developed infrastructure and offers a wide range of accommodation options.

**Pros**

- Vibrant nightlife and entertainment options
- Beautiful beaches and outdoor activities
- Proximity to Bangkok (easy access for weekend trips)

**Cons**

- Can be crowded and touristy
- Some areas have a reputation for seedy nightlife
- Higher cost of living in popular areas (e.g. Jomtien Beach)

**Neighbourhoods**

- **Jomtien**: Known for its quieter and more family-friendly atmosphere, Jomtien is a great place for those who want to enjoy the beach life without the hustle and bustle of central Pattaya. It offers a variety of restaurants, cafes, and water sports.
- **Naklua**: A more traditional Thai neighbourhood with a mix of local and expat residents. Naklua offers a more authentic living experience with local markets and temples, and a slower pace of life.
- **Pratumnak Hill**: Located between Pattaya and Jomtien, Pratumnak Hill is a popular area for expats due to its convenient location and stunning sea views. It offers a mix of luxury condos and affordable apartments.

**Suitable for**

- Young professionals
- Families
- Nightlife enthusiasts

**10. Phuket**

Phuket is Thailand's largest island and a popular destination for tourists and digital nomads. Known for its stunning beaches, vibrant nightlife and luxurious resorts, Phuket offers a unique blend of relaxation and excitement. The island has a well-developed infrastructure and plenty of amenities to cater to expats.

**Pros**

- Beautiful beaches and outdoor activities
- Vibrant nightlife and social scene
- Wide range of accommodation and schooling options

**Cons**

- High cost of living compared to other areas in Thailand
- Can be crowded and touristy, especially during the high season (December to March)
- Increasing traffic congestion from tourism and real estate development

**Neighbourhoods**

- **Cherngtalay**: Known for its beautiful Bangtao Beach, Cherngtalay is a popular area for expats looking for a more upscale and serene environment. It offers a range of luxury villas, resorts, and beachfront properties. The area is also home to the Laguna Phuket complex, which includes hotels, restaurants, and recreational facilities.
- **Patong and Kathu**: Known for its lively nightlife and beach scene, Patong is a great place for those who love to socialise and enjoy the beach life. It's a bustling area with many bars, clubs, and entertainment options. Kathu, located between Patong and Phuket Town, offers a more local experience while still being close to the main attractions. It's a convenient and affordable area to live in, with a mix of local amenities and expat-friendly facilities.

- **Rawai**: A quieter area with a more relaxed atmosphere. Rawai is popular among those who prefer a slower pace and beautiful natural surroundings. It's a great place for families and those who enjoy outdoor activities.

**Suitable for**

- Affluent young professionals
- Families
- Party and beach enthusiasts

**Conclusion**

Thailand offers an incredible variety of places for digital nomads to call home, each with its own unique charm and benefits. Whether you prefer the bustling city life of Bangkok, the laid-back atmosphere of Chiang Mai, the stunning beaches of Phuket, or the serene beauty of Koh Lanta, there's something for everyone. Consider your lifestyle, budget, and personal preferences when choosing your new home base. With the right location, you can enjoy a fulfilling and productive digital nomad experience in the Land of Smiles. Happy house-hunting!

## RENTING IN PARADISE

Finding the right place to stay in Thailand can be one of the most exciting parts of your journey. Whether you're looking for a short-term rental to get your feet wet or a long-term home to settle into, Thailand offers a variety of options to suit every need and budget. From bustling city apartments to tranquil beachside bungalows, the choices are endless. In this chapter, we'll guide you through securing short-term and long-term rentals, highlight the best neighbourhoods, and provide tips to help you make the most informed decision.

## Short-Term Rentals

Short-term rentals are ideal if you're new to Thailand and want to explore different areas before committing to a long-term lease. They offer flexibility and convenience, making them perfect for digital nomads who value adaptability.

### Airbnb

Airbnb is a popular choice for short-term stays, offering a wide range of apartments, condos, and houses. It's flexible and allows you to test different areas before committing to a longer lease. Airbnb hosts in Thailand are usually very hospitable and can provide valuable local insights. Prices can vary widely depending on the location and the type of accommodation, but you can expect to find options ranging from budget-friendly to luxurious. Additionally, Airbnb provides user reviews and ratings, which can help you make an informed decision based on the experiences of previous guests.

### Hotels and hostels

Ideal for very short stays or if you're just passing through, hotels and hostels in Thailand cater to all budgets. From budget hostels for backpackers to five-star luxury hotels, you'll find a variety of options to suit your needs. Many hostels also offer private rooms if you prefer a bit more privacy. Staying in a hotel or hostel gives you the chance to meet other travellers and exchange tips and experiences. Some hostels even organise social events and tours, providing a great way to explore the area and make new friends.

### Serviced apartments

Serviced apartments offer the comforts of home with the convenience of hotel services, making them perfect for stays ranging from a few weeks to a few months. These apartments typically come fully furnished and include amenities such as housekeeping, laundry services, and even on-site restaurants.

They provide a comfortable and hassle-free living experience, allowing you to focus on your work and exploration. Major cities like Bangkok, Chiang Mai, and Phuket have numerous serviced apartment options. Many serviced apartments also offer flexible lease terms, catering to the needs of digital nomads who may not want to commit to a long-term contract.

Websites to check out:

- **The Ascott Limited:** Offers a range of luxurious serviced apartments across Thailand, with properties in Bangkok, Pattaya, and other major cities.
- **Oakwood:** Provides high-quality serviced apartments in key locations such as Bangkok and Phuket, known for their excellent service and modern amenities.
- **Somerset Serviced Residence:** Features contemporary serviced apartments in Bangkok and other cities, ideal for short and extended stays.

**Vacation rentals**

Websites like **Booking.com** and **Agoda** also offer vacation rentals, which can be a good alternative to Airbnb. These platforms provide a wide range of options, from cosy apartments to beachfront villas, often with flexible booking terms. Vacation rentals can be particularly appealing if you want a unique or luxurious stay. These platforms also offer detailed descriptions, photos, and user reviews, making it easier to find a rental that matches your preferences and budget. You can also get cashback offers and earn points, which will be redeemable in the future.

**Long-Term Rentals**

Once you've settled on your preferred area, you might want to consider a long-term rental. Long-term rentals are more cost-effective and provide a sense of stability, which can be important for digital nomads who plan to stay in Thailand for several months or years.

**Apartments**

Apartments are more affordable than condos and are a good option for budget-conscious nomads. They may not have as many amenities, but they are still comfortable and convenient. Apartments are widely available in cities and towns across Thailand. Prices for apartments can range from **THB 10,000 to 30,000 per month**. Keep in mind that lower-priced apartments may not include furnishings, so you might need to budget for some basic furniture and appliances. Despite being more basic, many apartments are located in central areas with easy access to public transport, markets, and other amenities.

**Condominiums**

Condominiums are a popular choice among expats and digital nomads, often coming with amenities like pools, gyms, and security. Condos can be found in all major cities and offer a comfortable and modern living experience. They are especially popular in Bangkok, where high-rise condo buildings provide stunning city views. Renting a condo can range from **THB 20,000 to 50,000 per mont**h, depending on the location and amenities. Many will come with 24-hour security, parking, a gym, and a pool. Some condominiums offer additional amenities such as communal lounges, co-working spaces, and rooftop gardens.

**Houses**

If you prefer more space and privacy, renting a house might be the best option. Houses are plentiful in suburban areas and can offer a more local experience. You'll find various options, from traditional Thai houses to modern villas. Renting a house is ideal for families or those who enjoy gardening and outdoor space. Prices for houses typically range from **THB 30,000 to**

**100,000 per month**, depending on size and location. Houses often have private gardens, parking spaces, and sometimes even private pools, providing a homely and comfortable living environment.

### Townhouses

Townhouses, or 'townhomes', are another option for long-term rentals. These are typically multi-storey homes that share walls with neighbouring units. They offer more space than apartments and are often located in residential communities with shared facilities like parks and playgrounds. Townhouses can be a good compromise between a house and an apartment, offering space and community amenities. They are ideal for families or those who appreciate more space and a neighbourhood feel. Townhouses usually come with multiple bedrooms, making them suitable for shared living arrangements as well. Prices for townhouses typically range from **THB 15,000 to 40,000 per month**, depending on size and location.

### Conclusion

Finding the right rental in Thailand is a key step in making your digital nomad journey both comfortable and enjoyable. Whether you opt for the flexibility of short-term rentals or the stability of a long-term lease, there are plenty of options to suit your needs and budget. From the convenience of Airbnb and serviced apartments to the community feel of condos and houses, Thailand offers a diverse range of living arrangements. Take your time to explore different areas and types of accommodation to find the perfect fit for your lifestyle. Happy house-hunting and enjoy your stay in the Land of Smiles!

## FINDING THE PERFECT PLACE

Finding the perfect place to live in Thailand can be an exciting adventure. Whether you're looking for a short-term rental to explore different areas or a long-term home to settle into,

Thailand offers various options to suit every need and budget. With a bit of research and some local know-how, you can find a place that feels just right. Let's dive into some tips for house-hunting both online and on-site, as well as how to make the most of local resources. From bustling cities to serene beach towns, Thailand has it all.

## House-Hunting 101

### Online tips

In the internet age, finding a rental property online is a breeze. Here's how to make the most of it:

- **Use reputable platforms**: Websites like **DDproperty**, **Hipflat, Siam Real Estate**, **FazWaz**, **Thai Property**, **Livinginsider**, **RentHub**, and **Conrad Properties** are your go-to places. These sites let you filter properties by location, price, and type, making your search easy and efficient.
- **Check reviews and ratings**: Pay attention to reviews and ratings from previous tenants. They can give you the lowdown on the property condition and the landlord's reliability.
- **Join expat forums and Facebook groups:** Communities like **Thailand Expats**, **Bangkok Housing**, **and Rental and Property** on Facebook are treasure troves of information. Numerous online forums, such as **ASEAN Now**, **Pattaya Addicts**, **TeakDoor**, **Thaiger Talk**, and **ThailandQA**, allow you to ask pertinent questions and find answers from knowledgeable fellow expats. For a complete list, visit the **Resources** section at the back of this book.
- Fellow nomads and expats often share their experiences, recommend landlords, and give advice on the best neighbourhoods.

- **Request virtual tours and video calls**: If you can't visit in person, request a virtual tour or set up a video call with the landlord. This way, you get a better sense of the place before committing.
- **Use property search apps**: Apps like **Dot Property**, **PropertyScout**, and **Thailand Property** make it easy to search for rentals on the go. They offer user-friendly interfaces and up-to-date listings.

**On-site tips**

While online searches are handy, nothing beats seeing a place in person. Here's how to ace your on-site search:

- **Explore the neighbourhood**: Spend time walking around the areas you're interested in. This gives you a feel for the local vibe and amenities. You might even stumble upon rental signs that are not listed online.
- **Visit multiple properties**: Don't settle for the first place you see. Check out several properties to compare. Take notes and snap photos to keep track.
- **Check transport links and amenities**: Ensure the property is near public transport, grocery stores, hospitals, schools (if needed), fitness centres, and other essentials. This can make a huge difference in your daily life.
- **Talk to locals**: Chat with residents or local business owners. They can offer valuable insights into the area's safety and living conditions.
- **Attend open houses**: Open houses are great opportunities to view multiple properties quickly. They also allow you to meet landlords or property managers directly.

**Benefits of using real estate agents**

Real estate agents can be lifesavers in your house-hunting adventure. Here's why:

- **Local market knowledge**: Agents know the local market inside out. They can guide you to the best neighbourhoods, upcoming developments, and fair rental prices.
- **Language barrier**: Navigating rentals in a foreign country can be tricky due to language differences. A local agent can help bridge this gap, ensuring clear communication with landlords and handling paperwork.
- **Negotiation skills**: Skilled agents can negotiate better rental terms or lower prices on your behalf.
- **Timesaving**: Agents can save you time by shortlisting properties that match your criteria, arranging viewings, and taking care of paperwork.
- **No cost to you**: One of the best parts is that in Thailand, the renter doesn't pay the agent's commission – the landlord does. This means you can benefit from all the agent's expertise and services without it costing you a penny.

**Lease agreements**

Understanding lease agreements is key to a smooth rental experience. Here's what to look out for:

**Lease duration**: Most leases in Thailand are for six months to a year, but some landlords may agree to shorter terms, especially in tourist areas. Be clear about your needs and negotiate accordingly.

- **Deposits and payments**: Usually, you'll need to pay a security deposit (one to two months' rent) and the first month's rent upfront. Make sure you understand the terms for deposit return and any deductions.

- **Maintenance and utilities**: Clarify who's responsible for maintenance and utility bills. Some leases include utilities, while others don't. It's important to know what's covered to avoid surprises.
- **Walk-through inspection**: Do a thorough walk-through inspection before signing the lease. Document any existing damages with photos and notes. This can protect you from unfair charges when you move out.
- **Legal considerations**: Ensure the lease is in Thai and English if you're not fluent in Thai. If possible, have a legal expert review the contract to protect your interests.

**Furnishing your home**

Depending on your rental, you might need to furnish your new home. Here's how to go about it:

**Fully furnished vs. unfurnished**: Fully furnished rentals are convenient but may come with higher rent. Unfurnished places offer more flexibility in decorating but require an initial investment in furniture.

- **Shopping for furniture**: Budget-friendly options include **IKEA**, **HomePro**, **BnB Home**, **NocNoc**, and **SB Furniture**, which offer a wide range of furniture and home goods. For more unique, luxury or customised pieces, check out local furniture suppliers like **Index Living Mall**, **Chanintr**, **Chic Republic**, and **Euro Creations**.
- **Second-hand options**: Save money by buying second-hand furniture. **Facebook Marketplace** and local classified ads often have great deals on gently used items.
- **Renting furniture**: If you're staying short-term, consider renting furniture from companies like **Baania**

**Furnish**. This can be a cost-effective and hassle-free solution.

- **Local markets and shops**: Visit local markets and shops for affordable and unique home décor. These places often have handcrafted items that can add a personal touch to your home.
- **Appliances and tools**: if you need to fill your house with white goods and/or do some minor repairs at some point, check out these hardware stores: **HomePro**, **BnB Home**, **Tai Watsadu**, **Global House**, **PowerBuy**, and **DoHome**.

**Living in a community**

Settling into a new home also means becoming part of the local community. Here are some tips:

**Get to know your neighbours**: Building relationships with your neighbours can enhance your living experience. They can offer local insights, recommend services, and provide a sense of security.

- **Explore local amenities**: Take time to explore nearby markets, parks, and community centres. Participating in local events and activities can help you feel more connected and at home.
- **Join community groups**: Many neighbourhoods have community groups or social clubs. Joining these can provide opportunities to meet new people and engage in local culture.
- **Respect local customs**: Understanding and respecting local customs and etiquette can go a long way in building positive relationships within your community.
- **Volunteer and participate**: Volunteering for local causes or participating in community projects can be a great way to integrate and give back to your new home.

**Conclusion**

Finding the perfect place to live in Thailand is an exciting part of your digital nomad journey. By using both online resources and on-site visits and considering the help of local real estate agents, you can find a rental that suits your needs. Understanding lease agreements and furnishing your home properly will ensure a smooth transition. Finally, integrating into the local community will enrich your experience and make your stay in Thailand truly memorable. Happy house-hunting.

## COLLABORATIVE LIVING

Co-living spaces are becoming increasingly popular among digital nomads in Thailand. They offer a great way to connect with other like-minded individuals and provide a supportive environment for both living and working. These spaces combine the comforts of home with the productivity of an office, making them ideal for remote workers seeking a balanced lifestyle. Co-living spaces often have shared amenities like kitchens, lounges, and recreational areas, not to mention dedicated co-working spaces with high-speed internet, meeting rooms, and networking events. This combination fosters a sense of community and collaboration, making it easier to build professional and personal relationships while staying productive. For more about co-working spaces, please visit **Chapter 2: Work & Productivity**.

### Benefits of Co-Living Spaces

#### Community, networking, and cultural exchange

One of the biggest perks of co-living spaces is the sense of community they create. For digital nomads, who often travel solo and work remotely, co-living spaces offer an instant network of like-minded individuals. This community can be invaluable for both personal and professional growth. You'll have the chance to meet people from various backgrounds and industries, leading to new friendships, collaborations, and business opportunities.

Co-living spaces in Thailand also offer a unique opportunity for cultural exchange and personal growth. You'll immerse yourself in Thai culture while interacting with people from around the world, broadening your horizons and enriching your overall experience.

Many co-living places organise activities to help residents connect and share their skills and experiences. Whether it's a casual get-together over dinner, a skill-sharing session, a chess club meetup, a residents vs staff soccer match, or a professional networking event, these activities enhance your co-living experience. Furthermore, cultural activities and local experiences such as cooking classes, movie nights, language exchanges, or guided tours of nearby attractions deepen your connection to Thailand and make your stay more meaningful.

**Shared amenities and cost savings**

Another major benefit of co-living spaces is access to shared amenities. Most co-living spaces are fully furnished and equipped with everything you need, from high-speed internet and modern kitchens to laundry facilities and comfortable common areas. In many cases, you'll also have access to amenities that might be out of reach in a traditional rental, such as a gym, pool, or rooftop terrace.

By sharing these amenities with other residents, you can enjoy a higher standard of living without the associated costs. Co-living spaces often include utility bills, internet, cleaning, and other expenses in the monthly rent, making it easier to manage your budget. This all-inclusive pricing model can be particularly appealing for digital nomads, as it simplifies financial planning and reduces the stress of managing multiple bills.

**Flexibility and convenience**

Co-living spaces are designed with flexibility in mind, offering a range of rental options to suit different needs. Whether you're looking for a short-term stay of a few weeks or a long-term arrangement of several months, you can usually find a co-living space that fits your requirements. This flexibility is ideal for digital nomads, who may not want to commit to a long lease or might need to move frequently.

In addition to flexible rental terms, co-living spaces often provide a hassle-free living experience. Many spaces offer cleaning services, maintenance, and other support to ensure residents can focus on their work and personal pursuits rather than household chores. This convenience can greatly enhance your quality of life and free up more time to explore Thailand and enjoy your digital nomad lifestyle.

**Easy meal and grocery delivery**

Speaking of convenience, how about food delivered to your door! Living in Thailand also means accessing various convenient and cost-efficient meal delivery options. If you're not in the mood to cook, traditional delivery services from restaurants and cafes can save the day. Popular platforms such as **Grab**, **Foodpanda**, **Robinhood**, **Wongnai**, **Aroi Dee**, and **Eatigo** offer a wide range of cuisines delivered right to your door. You can explore local Thai dishes, international cuisines, and even some hidden gems from the comfort of your home.

For those who prefer healthier, pre-prepared meals, there are meal preparation services that deliver nutritious and delicious options. Companies like **Absotute FitFood**, **Pure Prep**, **Getfresh**, **Nutrichef**, **Healthy Meals**, **Paleo Robbie**, and **Vegan Crush** cater to different dietary needs, from paleo and keto to vegan and gluten-free. These services provide the convenience of ready-to-eat meals without the hassle of planning and cooking.

If you enjoy cooking but don't have time to grocery shop, there are also grocery and supermarket delivery services available. **HappyFresh**, **Lotus's**, **Tops**, **BigC**, **Makro**, and **Villa** offer a wide selection of groceries, fresh produce, and household items delivered straight to your doorstep. These services make it easy to maintain a well-stocked kitchen without navigating busy markets and stores.

Find more details about local eating options in **Chapter 8: Food & Dining**.

**Enhanced productivity and supportive environment**

For many digital nomads, maintaining productivity while working remotely is a top priority. Co-living spaces are often designed with this in mind, offering dedicated workspaces and high-speed internet to support professional needs. Many also have quiet areas, meeting rooms, and other facilities to help you stay focused and productive. Living among other remote workers adds motivation and inspiration, creating a supportive atmosphere to help you achieve your professional goals, whether working on a solo project or collaborating with fellow residents.

Moreover, co-living spaces provide a supportive environment that's particularly valuable for digital nomads. Moving to a new country and adjusting to a new way of life can be challenging, but co-living offers a built-in support network. From practical advice on navigating local customs and logistics to emotional support during difficult times, the community aspect can make a significant difference.

Many co-living spaces have staff members available 24/7 to assist residents with various needs, from arranging transportation to recommending local services. This level of support can make the transition to living in Thailand smoother and more enjoyable, ultimately enhancing both your productivity and overall experience.

## Top 20 Co-Living Spaces in Thailand

Co-living spaces provide a communal living environment where residents share amenities and common areas. They are designed to foster a sense of community and collaboration, making them perfect for digital nomads who thrive in social settings. Here are the best of the best, listed alphabetically.

### Alt Co-living (Chiang Mai)

Alt Co-living in Chiang Mai offers a modern and stylish living environment for digital nomads. The space features fully furnished rooms, shared kitchens, and communal lounges. Residents can take advantage of high-speed internet, meeting rooms, and regular community events. Alt Co-living is located close to popular cafes and co-working spaces, making it a convenient choice for remote workers.

### Amy Village (Koh Samui)

Amy Village offers a calming co-living space surrounded by lush greenery on Koh Samui. The facilities include fully furnished rooms, shared kitchens, and relaxing communal areas. High-speed internet, a swimming pool, and community events are some perks. It's perfect for digital nomads seeking a serene and productive environment.

### El Pillax Lanta Resort (Koh Lanta)

El Pillax Lanta Resort offers a luxurious co-living experience on the beautiful island of Koh Lanta. The resort features fully furnished rooms, shared kitchens, and stunning communal areas, including a beachfront lounge. Residents can enjoy high-speed internet, a dedicated co-working space, and frequent community events. With its picturesque surroundings and premium amenities, El Pillax Lanta Resort is perfect for digital nomads seeking comfort and productivity.

**HOMA (Phuket Town, Cherngtalay)**

HOMA offers co-living spaces in two prime locations in Phuket – Phuket Town and Cherngtalay. Designed with digital nomads in mind, HOMA features fully furnished apartments, co-working spaces, and communal facilities such as swimming pools, fitness centres, and rooftop gardens. Residents can enjoy high-speed internet, meeting rooms, and various social activities. HOMA's community events, fitness classes, and rooftop movie screenings make it easy to meet new people and stay productive.

**Note:** there is also a third branch located 40 minutes north of Pattaya in Si Racha. This development is more distant and has a Japanese décor style.

**Hub53 (Chiang Mai)**

Hub53 offers a unique co-living experience in Chiang Mai, combining modern living spaces with a vibrant co-working environment. The space features fully furnished rooms, shared kitchens, and communal areas. Residents can enjoy high-speed internet, meeting rooms, and a range of social activities. Hub53 is located close to popular cafes and co-working spaces, making it an ideal choice for digital nomads.

**KoHub (Koh Lanta)**

KoHub is a well-known co-living and co-working space in Koh Lanta. It offers comfortable accommodations, shared kitchens, and a vibrant co-working environment. Residents can enjoy high-speed internet, meeting rooms, and a range of community events. KoHub is located close to the beach, making it an ideal choice for those who want to combine work with leisure.

**LITA (Bangkok)**

LITA in Bangkok provides a stylish and comfortable living environment for digital nomads. The space features fully furnished rooms, shared kitchens, and communal lounges. Residents can enjoy high-speed internet, meeting rooms, and regular community events. LITA is known for its friendly and inclusive community, making it a welcoming place for newcomers.

**lyf Sukhumvit Soi 8 (Bangkok)**

lyf Sukhumvit Soi 8 offers a modern co-living experience in the heart of Bangkok. The space features fully furnished rooms, shared kitchens, and communal areas. Residents can enjoy high-speed internet, meeting rooms, and various social activities. lyf Sukhumvit Soi 8 is located close to popular cafes, restaurants, and co-working spaces, making it a convenient choice for remote workers.

**Mimi & Jim: Co-Living Space (Hua Him/Cha-am)**

Located just a few hours east of Bangkok on the outskirts of Hua Hin, Mimi & Jim offers a vibrant community for digital nomads and travellers. This modern, eco-friendly space features comfortable private rooms, collaborative coworking areas, and easy beach access. With fast Wi-Fi, regular social events, and a supportive atmosphere, it perfectly balances work and leisure. It's a haven for professionals seeking productivity and connection in a laidback and scenic coastal town.

**One Day (Bangkok)**

One Day in Bangkok offers a modern and stylish co-living environment for digital nomads. The space features fully furnished rooms, shared kitchens, and communal areas. Residents can enjoy high-speed internet, meeting rooms, and a range of social activities. One Day is located close to popular cafes, restaurants, and co-working spaces, making it a convenient choice for remote workers.

**Roost Glamping (Koh Lanta)**

Roost Glamping offers a unique co-living experience with a focus on outdoor living. The space features glamping tents, shared kitchens, and communal areas. Residents can enjoy high-speed internet, a co-working area, and regular community events. Roost Glamping is located close to the beach, making it an ideal choice for those who enjoy nature and outdoor activities.

**Secret Place Hotel & Coworking (Koh Phangan)**

Secret Place Hotel & Coworking provides a serene and productive environment for digital nomads in Koh Phangan. The space includes fully furnished rooms, a well-equipped kitchen, and various communal areas. Residents benefit from high-speed internet, a professional co-working space, and a host of social activities such as yoga sessions and community dinners. It's an ideal choice for those who value both work and relaxation in a tropical setting.

**Selina – Serenity Rawai (Phuket)**

Selina – Serenity Rawai in Phuket offers a beachfront co-living experience for digital nomads. The space features fully furnished rooms, shared kitchens, and communal areas. Residents can enjoy high-speed internet, meeting rooms, and a range of social activities. Selina – Serenity Rawai is located close to popular beaches, cafes, and co-working spaces, making it an ideal choice for remote workers.

**Socialer (Chiang Mai)**

Socialer in Chiang Mai offers a vibrant and inclusive co-living environment designed for digital nomads and remote workers. The space provides comfortable, fully furnished rooms, shared kitchen facilities, and inspiring communal areas. Residents enjoy high-speed internet, access to co-working spaces, and a range of community-focused events aimed at fostering connections and collaboration. Located near popular cafes and cultural spots, Socialer is perfect for those who thrive in a dynamic community.

**TAO HUB (Koh Tao)**

TAO HUB on Koh Tao is known for offering a co-living space with stunning ocean views. The space typically features fully furnished rooms, shared kitchens, and a spacious co-working area. Residents can enjoy high-speed internet, meeting rooms, bike rentals, and beachfront activities. TAO HUB is perfect for those who want to combine work with diving and snorkelling.

**The Cove (Koh Phangan)**

The Cove is a beachfront co-living space on the island of Koh Phangan. It offers a range of accommodation options, from dormitory-style rooms to private bungalows. The co-working area is equipped with high-speed internet, comfortable workstations, and meeting rooms. The Cove also features a beachfront cafe, yoga studio, and communal kitchen. Regular social events, workshops, and networking sessions help residents connect and collaborate.

**The Social Club (Chiang Mai)**

The Social Club offers a unique co-living experience in Chiang Mai, combining modern living spaces with a vibrant co-working environment. The space features fully furnished rooms, shared kitchens, and communal areas. Residents can enjoy high-speed internet, meeting rooms, and various social activities. The Social Club is located close to popular cafes and co-working spaces, making it an ideal choice for digital nomads.

**The Yard Hostel (Bangkok)**

The Yard Hostel in Bangkok offers a relaxed co-living environment focusing on community. The space features comfortable dormitory-style rooms, shared kitchens, and communal areas. Residents can enjoy high-speed internet, a co-working area, and regular social events. The Yard Hostel is known for its friendly atmosphere and inclusive community.

**Ulima Co-living Hotel (Bangkok)**

Ulima Co-living Hotel in Bangkok offers a unique blend of amenities and co-living spaces. The space features fully furnished rooms, shared kitchens, and communal lounges. Residents can enjoy high-speed internet, meeting rooms, and regular community events. Ulima Co-living Hotel is located close to popular attractions and co-working spaces, making it an ideal choice for digital nomads.

**UnionSPACE (Bangkok)**

UnionSPACE in Bangkok provides a stylish and comfortable living environment for digital nomads. The space features fully furnished rooms, shared kitchens, and communal lounges. Residents can enjoy high-speed internet, meeting rooms, and regular community events. UnionSPACE is known for its friendly and inclusive community, making it a welcoming place for newcomers.

**Note**: Things change constantly in the Land of Smiles. Once you settle somewhere, be sure to do a quick internet search to see what new places might have popped up since this book was published. You never know what new gems you might discover!

**Conclusion**

Co-living spaces in Thailand offer a unique blend of community, convenience, and comfort, making them an excellent choice for digital nomads. From the bustling streets of Bangkok to the serene beaches of Phuket, there's a co-living option to suit every lifestyle and preference. By choosing a co-living space, you can enjoy the benefits of shared amenities, vibrant social scenes, and supportive environments, all while staying productive and connected. As you explore these options, you'll find that co-living can greatly enhance your experience as a digital nomad in Thailand.

**Stay tuned . . .**

Now that you've got a handle on where to live, it's time to focus on your work environment. In the next chapter, we'll dive into everything you need to know about working remotely in Thailand. From finding the best co-working spaces to ensuring reliable internet connectivity and balancing work with leisure, we've got you covered. Whether you're settling into a bustling city or a tranquil beach town, having the right work setup is crucial for maintaining productivity and achieving a healthy work-life balance. We'll provide insights into the most popular co-working hubs, tips on staying connected, and advice on integrating work with the rich cultural experiences Thailand has to offer.

Let's get ready to boost your productivity and maximise your time in the Land of Smiles! Don't miss **Chapter 2: Work & Productivity** for comprehensive tips and strategies to help you thrive in your remote work environment. From practical advice on setting up your workspace to suggestions for networking with fellow professionals, this chapter will equip you with the tools you need to succeed. Happy house-hunting and enjoy your stay in the Land of Smiles!

# Chapter 2:
# Work & Productivity

## Getting Stuff Done: Working Remotely in Thailand

In an age where work is no longer confined by office walls, Thailand has emerged as a top destination for digital nomads seeking a harmonious blend of productivity and adventure. Whether drafting reports from a beachfront café, attending virtual meetings in a bustling Bangkok co-working space, or coding from a serene jungle retreat, Thailand offers diverse environments to suit every professional's needs. This chapter guides you through maintaining productivity while enjoying Thailand's unique lifestyle, from finding the perfect co-working space to ensuring reliable internet connectivity. Your journey to becoming a productive digital nomad in Thailand starts here.

## CO-WORKING SPACES

Co-working spaces provide a professional and productive environment for remote workers, making them an ideal choice for digital nomads in Thailand. These spaces are designed to facilitate collaboration and networking, offering a range of amenities to support your work needs. Whether you prefer a quiet corner to focus on your tasks or a vibrant communal area to brainstorm with like-minded professionals, co-working spaces cater to diverse work styles. Many come equipped with high-speed internet, meeting rooms, ergonomic furniture, and even recreational areas to help you unwind. With options available in major cities and popular tourist destinations, you'll always find a suitable spot to stay productive.

## Introduction to Co-Working Spaces

Whether you're a freelancer, entrepreneur, or part of a remote team, co-working spaces provide the infrastructure and support needed to thrive in a dynamic work environment. Let's dive in and look at them in more detail.

### Proc and cons of co-working

**Pros**

- **Professional environment**: Provides a structured and professional setting, enhancing productivity.
- **Networking opportunities**: Facilitates connections with other professionals, leading to potential collaborations.
- **Amenities**: Offers a range of amenities such as high-speed internet, meeting rooms, and recreational areas.
- **Flexibility**: Various membership options allow for flexibility in working arrangements.
- **Community**: Creates a supportive community that can provide social and professional support.

**Cons**

- **Cost**: Membership fees can be expensive, especially in premium locations.
- **Distractions**: Communal areas can sometimes be noisy and distracting.
- **Lack of privacy**: Shared spaces may lack the privacy needed for confidential work.
- **Dependence on location**: Quality and availability of co-working spaces vary by location.
- **Availability**: Popular co-working spaces can get crowded, making it difficult to find a spot.

### What to expect and how to choose the right one

Choosing the right co-working space can significantly impact your productivity and overall experience as a digital nomad.

Here are some key factors to consider when selecting a co-working space in Thailand.

**Location, location, location**

The location of your co-working space can greatly influence your daily routine and work-life balance. Consider spaces conveniently located near your accommodation, public transport, and essential amenities such as cafés, restaurants, and gyms. In bustling cities like Bangkok and Chiang Mai, you'll find co-working spaces in trendy neighbourhoods, offering easy access to a vibrant social scene. If you prefer a more relaxed environment, look for co-working spaces in coastal towns or quieter areas where you can enjoy nature and tranquillity.

**Amenities and facilities**

Different co-working spaces offer varied amenities, so it's essential to identify what you need to stay productive and comfortable. Common amenities include high-speed internet, air conditioning, ergonomic furniture, and access to meeting rooms. Some spaces also provide additional perks such as free coffee and snacks, printing services, on-site assistance, and lockers. If you anticipate needing a quiet space for calls or video conferences, ensure the co-working space has private rooms or soundproof booths. Recreational facilities like lounges, game rooms, and even on-site gyms can also enhance your overall experience.

**Community and networking opportunities**

One of the main advantages of co-working spaces is the opportunity to connect with other professionals. A strong community can provide valuable networking opportunities, collaboration possibilities, and social support. Look for spaces that host regular events, workshops, and networking sessions to help you engage with fellow digital nomads and local professionals. Some co-working spaces even have dedicated community managers who organise activities and facilitate connections among members. Joining a space with a dynamic and supportive community can enrich your experience and open doors to new opportunities. You never know; maybe the person beside you is looking to buddy up on a project.

**Flexibility and pricing**

Co-working spaces in Thailand offer various membership options to suit different needs and budgets. Consider whether you need a dedicated desk, a hot desk, or a private office. Hot desks are usually the most affordable option, allowing you to work from any available spot in the common area. In contrast, dedicated desks and private offices provide more privacy and stability. Many co-working spaces offer daily, weekly, and monthly passes, allowing you to choose a plan that fits your schedule. Be sure to compare prices and packages to find a co-working space that offers good value for money.

**Reviews and recommendations**

Before committing to a co-working space, it's a good idea to read reviews and seek recommendations from other digital nomads. Online platforms like **Google Reviews**, **Facebook Groups**, and **specialised forums (e.g. Reddit)** can provide insights into the experiences of other users. Pay attention to comments about the reliability of the internet, the friendliness of the staff, and the overall atmosphere. Additionally, consider contacting fellow digital nomads through online communities and social media groups for personal recommendations and tips.

**Trial periods and day passes**

Many co-working spaces offer trial periods or day passes, allowing you to experience the environment before making a long-term commitment. Take advantage of these options to test the facilities, assess the internet speed, and get a feel for the community. This can help you make an informed decision and ensure the space meets your needs and preferences. Some passes even come with additional 'perks' like access to an on-site lounge, gym, swimming pool and free event access.

**Accessibility and security**

Consider the accessibility and security of the co-working space. Ensure that the space is easily accessible by public transport or has adequate parking if you plan to drive. Check if the space operates 24/7 or has specific opening hours that align with your work schedule. Security features such as keycard access, CCTV, and secure storage for your belongings are also important to ensure a safe and comfortable working environment. Also, check how services can be accessed, for example, an app or WhatsApp contact to book private meeting rooms or soundproof pods when you're not on-site.

**Ambience and work environment**

The ambience and overall work environment play a crucial role in your productivity and comfort. Visit the co-working space to assess the noise levels, lighting, temperature, and overall atmosphere. Some spaces have a more corporate feel, while others are designed to be more relaxed and creative. Choose a space that aligns with your work style and preferences. Natural light, comfortable seating, and a pleasant environment can significantly enhance your work experience.

**Tech support and infrastructure**

Reliable tech support and robust infrastructure are essential for digital nomads who rely heavily on technology. Check if the co-working space provides on-site IT support to assist with any technical issues you might encounter. Additionally, ensure the space has backup power supplies and a stable internet connection to avoid disruptions during power outages or network failures.

### Health and safety measures

In the current global scenario, health and safety measures are more important than ever. Assess the co-working space's adherence to cleanliness and sanitation standards. Look for spaces that provide hand sanitiser stations, maintain regular cleaning schedules, and encourage social distancing. Ensuring a safe and hygienic environment can improve your peace of mind and overall well-being.

## Conclusion

Choosing the right co-working space is a crucial step in ensuring your productivity and overall satisfaction as a digital nomad in Thailand. By considering factors such as location, amenities, community, flexibility, and personal preferences, you can find a space that perfectly suits your work style and needs. Co-working spaces offer not just a place to work but also opportunities for networking, collaboration, and social interaction, enriching your experience as a digital nomad. With a myriad of options available in bustling cities and serene coastal towns alike, there's a co-working space in Thailand waiting to help you get your best work done. **Happy co-working!**

**Note: For more information on forums and other resources, be sure to check out the Resources section at the end of this book.**

## BEST CO-WORKING SPACES ACROSS THAILAND

Thailand offers a diverse array of co-working spaces that cater to the needs of digital nomads and remote workers. From bustling urban centres to serene coastal towns, these spaces provide not just a place to work but also a community to belong to. Each co-working space has its unique vibe, amenities, and community spirit, making it essential to find one that aligns with your work style and personal preferences. In this section, we will explore some of the best co-working spaces across the country, highlighting their key features and what makes them stand out.

## Top 20 Nationwide

### Beachub (Koh Phangan)

Beachub is a beachfront co-working space on the island of Koh Phangan. It offers high-speed internet, comfortable workstations, and meeting rooms. The space also features a beachfront café, lounge areas, and outdoor seating. Beachub hosts regular social events, workshops, and networking sessions to encourage collaboration and community-building among members.

### CocoVille (Phuket)

CocoVille in Phuket offers a modern co-working environment focusing on community and productivity. The space features high-speed internet, comfortable workstations, and meeting rooms. CocoVille also provides a range of amenities, including a café, lounge areas, and outdoor seating. The space hosts regular social events, workshops, and networking sessions to foster a sense of community among members.

### Garage Society (Phuket and Hua Hin)

The Garage Society combines modern co-working facilities with a strong community focus. With locations in Hua Hin and Phuket, the space features high-speed internet, ergonomic workstations, and meeting rooms. Members can take advantage of networking events, skill-building workshops, and social activities. The Garage Society's supportive environment and vibrant community make it an excellent choice for remote workers.

**Glowfish (Bangkok)**

Glowfish in Bangkok is a dynamic coworking space designed for innovators and businesses. Offering a variety of flexible workspaces, including private offices, hot desks, and meeting rooms, Glowfish enhances productivity with modern amenities and high-speed internet. With its vibrant community, regular events, and wellness facilities (hello, nap room!), this stylish space fosters creativity and collaboration. Conveniently located in the city's heart, this place is the perfect blend of functionality and inspiration for professionals.

**Greenhouse (Chiang Mai)**

Greenhouse in Chiang Mai offers a vibrant co-working environment designed to support productivity and creativity. The space features high-speed internet, comfortable workstations, and meeting rooms. Greenhouse also provides a range of amenities, including a café, lounge areas, and outdoor seating. The space hosts regular social events, workshops, and networking sessions to foster a sense of community among members.

**Grind Time Coworking Space (Phuket)**

Grind Time offers a professional and productive co-working environment in Chalong Bay and Rawai Beach, Phuket. The spaces feature high-speed internet, comfortable workstations, and meeting rooms. Grind Time also provides a range of amenities, including cafés, lounge areas, and outdoor seating. The spaces host regular social events, workshops, and networking sessions to foster a sense of community among members on the island.

**HIVE Coworking Space (Pai)**

Nestled in the picturesque town of Pai, HIVE Coworking Space offers a serene and creative environment for remote workers. The space features high-speed internet, comfortable workstations, and meeting rooms, making it an ideal spot for productivity. HIVE also boasts a beautiful garden area where you can work amidst nature. The co-working space hosts regular social events, workshops, and networking sessions to foster a sense of community among members. Its tranquil setting and supportive community make HIVE a hidden gem for digital nomads seeking a peaceful yet productive workspace.

**Hub53 (Chiang Mai)**

Hub53 offers a vibrant co-working environment in Chiang Mai designed to support productivity and collaboration. The space features high-speed internet, comfortable workstations, and meeting rooms. Hub53 hosts regular networking events, workshops, and social activities to help members connect and collaborate. The supportive community and modern facilities make Hub53 an ideal place for remote workers.

**Hubba-to (Bangkok)**

Located in the trendy suburb of Onnut (Habito Mall) in central Bangkok, Hubba-to is known for its modern facilities, and active community events. The co-working space is equipped with high-speed internet, ergonomic furniture, and meeting rooms. Hubba-to hosts various events, from networking sessions to skill-building workshops, making it easy to connect with other professionals. The lively community and dynamic environment make Hubba-to a popular choice for digital nomads.

**Kohub (Koh Lanta)**

Kohub is a beachfront co-working space on the island of Koh Lanta. It offers high-speed internet, comfortable workstations, and meeting rooms. The space also features a beachfront café, lounge areas, and outdoor seating. Kohub hosts regular social events, workshops, and networking sessions to encourage collaboration and community-building among members.

**Koh Mak Campus (Koh Mak)**

Koh Mak Campus offers a blend of co-working on the serene island of Koh Mak. It features high-speed internet, comfortable workstations, and meeting rooms. The venue offers a well-furnished work area, communal spaces, and organised social events, making it an ideal spot for remote workers seeking a balanced lifestyle.

**Let's Work (Phuket)**

Let's Work in Phuket provides a professional and productive co-working environment. The space features high-speed internet, comfortable workstations, and meeting rooms. It also offers amenities like a café, lounge areas, and outdoor seating. Regular social events, workshops, and networking sessions are held to foster community building among members.

**Oasis (Koh Chang)**

Oasis offers excellent co-working facilities on the beautiful island of Koh Chang. With high-speed internet, ergonomic workstations, and a variety of meeting rooms, it provides a conducive environment for productivity. The space includes comfortable working spaces, communal areas, and social events to foster community engagement.

**Punspace (Chiang Mai)**

A favourite among digital nomads in Chiang Mai, Punspace offers multiple locations (Wiang Kaew and Tha Phae Gate) with great amenities and a strong community vibe. Each location features high-speed internet, comfortable workstations, and meeting rooms. Punspace hosts regular networking events, workshops, and social activities to help members connect and collaborate. The vibrant community and supportive environment make it an ideal place for remote workers.

**rPod Coworking Space (Pattaya)**

Located in Pattaya, rPod provides a professional co-working environment complete with high-speed internet, ergonomic workstations, and various meeting rooms. The space also includes a café, lounge areas, and outdoor seating. Regular events such as workshops and networking sessions are held to encourage community interaction. There are well known locally for their regular **Meetup** events.

**Seedspace (Chiang Mai)**

Located in the cultural heart of Chiang Mai, Seedspace offers a modern and inspiring co-working environment. The space features high-speed internet, ergonomic workstations, and various meeting rooms. Seedspace is part of the Seedstars network, which means you'll have access to a global community of entrepreneurs and innovators. The co-working space also provides amenities like a café, lounge areas, and event spaces. Regular workshops, networking events, and pitch competitions make Seedspace a vibrant hub for digital nomads and local professionals alike.

**The Hive (Bangkok)**

The Hive, located in trendy Thonglor, is a premier coworking space catering to creative professionals and entrepreneurs. This stylish, versatile workspace offers private offices, hot desks, and meeting rooms, all equipped with high-speed internet. Members benefit from a dynamic community, networking events, and wellness programs. With its sleek design, rooftop terrace, and central location, The Hive provides an inspiring and productive environment for those looking to thrive in Thailand's bustling capital.

**The Work Loft (Bangkok)**

Located in Silom, one of Bangkok's bustling business districts, The Work Loft offers a well-equipped co-working space with high-speed internet, comfortable workstations, and meeting rooms. The space also includes amenities like a rooftop terrace, café, and lounge areas. Regular events and workshops provide ample networking opportunities in a professional yet relaxed environment. Being in the heart of the city, there are no shortage of places to go when you knock off work for the day.

**UnionSPACE (Koh Chang)**

UnionSPACE Koh Chang is a premier coworking destination nestled on the stunning island of Koh Chang, Thailand. This innovative space offers high-speed internet, modern workstations, and private meeting rooms. Its unique island location provides a serene and inspiring environment perfect for productivity and creativity. Regular community events and workshops foster collaboration and networking among members. Whether you're a digital nomad or entrepreneur, UnionSPACE Koh Chang offers an exceptional work-life balance.

**Yellow (Chiang Mai)**

Yellow is an innovative coworking space in Chiang Mai, recognised for its vibrant and collaborative atmosphere. Catering to remote workers and entrepreneurs, it boasts high-speed internet, ergonomic furniture, and versatile meeting rooms. Unique to Yellow is its comprehensive support for startups, including regular networking events, workshops, and social activities designed to foster connections. As a key player in one of Thailand's most dynamic startup ecosystems, Yellow provides a professional, productive, and community-focused environment, making it a top choice for digital nomads in the north.

**Note: Things change constantly in the Land of Smiles. Once you settle somewhere, be sure to do a quick internet search to see what new places might have popped up since this**

**book was published. You never know what new gems you might discover!**

## Conclusion

Choosing the right co-working space is a crucial step in ensuring your productivity and overall satisfaction as a digital nomad in Thailand. By considering factors such as location, amenities, community, flexibility, and personal preferences, you can find a space that perfectly suits your work style and needs. With a myriad of options available in bustling cities and serene coastal towns alike, there's a co-working space in Thailand waiting to help you get your best work done. Enjoy the unique blend of professionalism and paradise that Thailand offers and make the most of your remote working journey.

# INTERNET CONNECTIVITY

Ensuring reliable internet connectivity is one of the most important factors for digital nomads who rely on a stable connection to perform their work efficiently. Thailand, with its modern infrastructure and widespread internet access, is generally well-equipped to meet the needs of remote workers. However, knowing where to find the best connectivity and how to stay connected on the go can make a significant difference to your productivity and overall experience.

## Getting Online

When it comes to working remotely, a reliable internet connection is non-negotiable. Here are some strategies to ensure you have the best possible connectivity.

Ensuring reliable internet for work

### Co-working spaces and cafés

As discussed in the previous section, co-working spaces are a great option for reliable internet. Most co-working spaces in

Thailand offer high-speed internet with stable connections, making them ideal for video conferencing, uploading large files, and other bandwidth-intensive tasks. Many cafés also provide free wi-fi, but the speed and reliability vary greatly. It's always a good idea to test the connection before settling in for a long work session. Look for cafés that are popular among digital nomads, as they are more likely to have suitable internet facilities.

**Accommodation with strong wi-fi**

When choosing accommodation, whether it's a hotel, guesthouse, or rental apartment, always check the quality of the internet connection. Many listings on platforms like **Airbnb** and **Booking.com** include wi-fi information, and you can read reviews to see what previous guests have said about internet speed and reliability. If you're staying long-term, you might consider visiting potential accommodations in person to test the wi-fi before committing.

**Internet on the move**

Staying connected while on the go is crucial for digital nomads. Consider getting a local SIM card with a generous data plan from providers like **AIS**, **TrueMove**, and **DTAC** to use your smartphone as a hotspot. Portable wi-fi devices, often called 'pocket wi-fi', are another excellent option, offering high-speed internet wherever you are. Many public places in Thailand, such as shopping malls and airports, also offer free wi-fi hotspots, but always use a VPN to protect your data when using public networks.

Whilst it's rare these days to rent a place that doesn't have a connected internet line already installed, you might find yourself in a situation where there is no line installed, it has been disconnected, or you want to upgrade or change provider. Here is what you'll need to know.

**Choosing a provider**

Thailand has several reputable internet service providers (ISPs), such as **AIS, TrueMove, DTAC, TOT**, and **3BB**. Research their plans and packages to find the one that best suits your needs regarding speed, data limits, and cost. Most ISPs offer a range of options, from basic plans to high-speed fibre optic connections.

**Visit a service centre or apply online**

Once you've chosen a provider, you can either visit one of their service centres or apply online through their website. If you prefer in-person assistance, the service centres can be found in major malls and shopping centres. Bring necessary documents such as your passport, visa, and proof of address (rental agreement or utility bill).

**Installation process**

After you've selected a plan and submitted your application, the ISP will schedule an installation appointment. The installation process typically takes a few hours and involves setting up the modem/router, picking a secure password, and ensuring the connection is stable. Most ISPs offer free installation as part of the package, but it's always good to confirm this beforehand.

**Contract duration and costs**

Internet contracts in Thailand usually range from 12 to 24 months. Monthly costs vary depending on the speed and type of connection, with basic plans starting around **THB 500 per month** and high-speed fibre optic plans costing upwards of **THB 1,000 per month**. Some providers may also offer month-to-month plans but at a higher cost per month. Be sure to read the contract terms carefully, including any early termination fees.

### Upgrading or changing providers

If you already have an internet connection but wish to upgrade or switch providers, the process is relatively straightforward. Contact your current provider to discuss upgrade options or cancellation procedures. If switching, you'll need to follow the same steps as installing a new connection with your chosen new provider. Ensure any existing contract obligations are fulfilled to avoid penalties.

### Mobile internet and hotspots

Having a backup plan is essential for ensuring uninterrupted connectivity. Thailand's mobile internet infrastructure is robust, with 4G and 5G networks widely available in most urban and tourist areas. Major providers like **AIS**, **TrueMove**, **DTAC**, **TOT**, and **3BB** offer various data plans that cater to your needs. Consider getting a local SIM card with a generous data plan (i.e. unlimited) and use your smartphone as a hotspot in case your primary connection fails. Portable wi-fi devices, often called 'pocket wi-fi', are another excellent option for staying connected on the go. These devices can be rented from numerous providers and offer high-speed internet wherever you are.

### Packages and Pricing Plans

### The big 5

**Let's take a look at what mobile SIM packages the big five providers have on offer.**

### AIS

AIS is one of the most popular mobile providers in Thailand, known for its broad coverage and reliable service. Let's look at some of their data plans.

- **Traveller SIM THB 299**: Offers 3 GB of high-speed data for 8 days. Ideal for short stays.

- **Traveller SIM THB 599**: Provides 15 GB of high-speed data for 15 days. A good option for longer trips.
- **Monthly Unlimited Data Plan**: For around **THB 899 per month**, you get unlimited data with high-speed access up to a certain limit (usually 30 GB), after which the speed is reduced but remains usable.

**TrueMove**

TrueMove is another major provider with excellent coverage and competitive pricing. They offer a variety of plans to suit different needs.

- **Traveller SIM THB 299**: Includes 8 GB of high-speed data for 15 days. Suitable for short-term visitors.
- **Traveller SIM THB 599**: Comes with 30 GB of high-speed data for 30 days. Great for digital nomads who need more data.
- **Monthly Unlimited Data Plan**: For around **THB 799 per month**, you get unlimited data with high-speed access up to 60 GB, after which the speed is reduced.

**DTAC**

DTAC is well-regarded for its customer service and network reliability. Here are some of their popular data plans.

- **Happy Tourist SIM THB 299**: Provides 8 GB of high-speed data for 15 days. Ideal for tourists and short-term visitors.
- **Happy Tourist SIM THB 599**: Offers 15 GB of high-speed data for 30 days. A good choice for longer stays.
- **Monthly Unlimited Data Plan**: For around **THB 899 per month**, you get unlimited data with high-speed access up to 40 GB, followed by reduced speeds.

### TOT

TOT is a reputable provider known for its reliable service and extensive network coverage. Here are some of their data plans:

- **Traveller SIM THB 300**: Offers 10 GB of high-speed data for 10 days. Ideal for travellers staying for a short period.
- **Traveller SIM THB 600**: Provides 20 GB of high-speed data for 20 days. Suitable for longer visits.
- **Monthly Unlimited Data Plan**: For around **THB 850 per month**, you get unlimited data with high-speed access up to 50 GB, after which the speed is reduced.

### 3BB

3BB is another trusted provider offering a range of data plans to meet various needs. Here are some of their popular packages:

- **Traveller SIM THB 250**: Includes 5 GB of high-speed data for 10 days. Great for short-term visitors.
- **Traveller SIM THB 550**: Comes with 25 GB of high-speed data for 30 days. Ideal for digital nomads who need more data.
- **Monthly Unlimited Data Plan**: For around **THB 800 per month**, you get unlimited data with high-speed access up to 70 GB, after which the speed is reduced.

### Choosing the right plan

When selecting a mobile data plan, consider the following factors.

- **Duration of stay**: Short-term visitors might prefer the **Traveller** or **Tourist SIMs**, while long-term digital nomads may find the monthly unlimited plans more suitable.
- **Data usage**: Estimate the data you'll need based on your typical usage patterns. If you rely heavily on video calls,

streaming, gaming, or large file uploads, opt for plans with higher data limits.
- **Coverage**: While all three providers offer extensive coverage, you might want to check specific coverage maps if you plan to stay in more remote areas.

**How to get a SIM card**

Getting a SIM card in Thailand is straightforward.

- **At the airport**: Major airports in Thailand have kiosks for **AIS**, **TrueMove**, and **DTAC**, where you can purchase SIM cards upon arrival.
- **Retail stores**: Visit any of the numerous retail stores or authorised dealers throughout the country.
- **Online**: Some providers offer online purchases and even delivery services to your accommodation.

**Activation and top-up**

Activating your SIM card is usually quick and can be done at the point of purchase. Top-ups can be done online, at convenience stores like **7-Eleven**, or via the provider's mobile app.

**Internet cafés**

In areas where co-working spaces and reliable cafés are scarce, internet cafés can be a viable alternative. These establishments often provide decent internet connections and a conducive environment for work. While they might not offer the same level of comfort or amenities as co-working spaces, they can be a good fallback option when needed. The only thing to remember is that they also don't offer the same level of privacy or noise reduction (you'll likely be sitting beside a noisy gamer).

**Fibre optic connections**

If you're planning to stay in Thailand for an extended period and require a highly stable connection, consider installing a fibre optic internet connection in your accommodation. Many urban areas in Thailand offer fibre optic internet, which provides faster and more reliable connectivity than traditional broadband. This option might be more expensive and require a longer commitment, but it can be worth it for those who need the best possible internet for their work.

**Staying Connected on the Go**

Being a digital nomad often means moving around and working from various locations. Below are some tips to help you stay connected while on the move:

**Download offline resources**

When travelling to areas with potentially unreliable internet, it's wise to download essential resources in advance. This includes documents, emails, and any other materials you might need for work. Many cloud storage services like **Google Drive** and **Dropbox** offer offline access options, allowing you to work without an internet connection and sync your changes later.

**Use internet mapping tools**

Tools like **ADSL Thailand Speedtest**, **Speedtest by Ookla** and **WiFi Map** can help you find locations with strong internet connections. Speedtest allows you to check the speed of your current connection, while WiFi Map offers a database of wi-fi passwords for public networks worldwide. These tools can be invaluable when you're in a new area and need to find a reliable internet connection quickly.

**Invest in a good VPN**

A virtual private network (VPN) is crucial for maintaining online security, especially when using public wi-fi networks. A

VPN encrypts your internet connection, protecting your data from potential hackers. Additionally, a VPN can help you access geo-restricted content, which can be useful if you need to connect to services or websites unavailable in Thailand. Popular VPN services like **ExpressVPN**, **NordVPN**, and **CyberGhost** offer reliable and secure connections.

**Keep your devices charged**

Power outages can occur, especially in more remote areas. Always keep your devices charged and consider investing in a portable power bank. This ensures you can continue working even if you temporarily lose access to a power source. Some co-working spaces and cafés also offer charging stations, so take advantage of these facilities whenever possible.

**Plan for time zone differences**

Thailand operates in the Indochina Time Zone (ICT), seven hours ahead of Coordinated Universal Time (UTC+7). If you're working with clients or colleagues in different time zones, plan your schedule accordingly to ensure seamless communication and collaboration. Use tools like **World Time Buddy** or **Google Calendar** to manage your time effectively and avoid missed meetings or deadlines.

**Join online communities**

Online communities and forums can be a great resource for staying updated on the best places for internet connectivity. Platforms like **Nomads.com**, **LinkedIn**, **Reddit**, and various **Facebook Groups** dedicated to digital nomads in Thailand offer valuable insights and recommendations from fellow remote workers. Engaging with these communities can help you discover hidden gems and avoid potential connectivity pitfalls.

**Test your connection regularly**

Regularly testing your internet connection can help you identify and address potential issues before they become problematic.

Use tools like Speedtest to check your connection speed and stability. If you notice a significant drop in performance, consider switching to a different network or location to maintain your productivity.

**Conclusion**

Ensuring reliable internet connectivity is a critical aspect of successfully working remotely in Thailand. By carefully selecting your co-working space, accommodation, and mobile data plans, you can maintain a stable connection that supports your professional needs. Additionally, adopting strategies for staying connected on the go, such as downloading offline resources and using VPNs, will help you navigate potential connectivity challenges. With some preparation and the right tools, you can enjoy the flexibility and freedom of being a digital nomad in Thailand while staying productive and connected.

# BALANCING WORK & PLAY

One of the most appealing aspects of being a digital nomad in Thailand is the ability to blend work with the pleasures of travel and leisure. The country's beautiful landscapes, vibrant culture, and welcoming community create an environment where you can enjoy life to the fullest while meeting your professional obligations. However, balancing work and play can be challenging, especially when so many tempting distractions surround you. This section provides practical tips on time management and scheduling to help you strike the perfect balance between productivity and enjoyment.

## Staying on Top of Things

Effective time management is crucial for maintaining productivity while enjoying all that Thailand has to offer. Here are some strategies to help you stay on top of your work without sacrificing leisure time.

## Time management tips

### Set clear goals and priorities

Before diving into your workday, take a moment to set clear goals and priorities. Determine which tasks are most important and need immediate attention and which can be deferred. Tools like the **Eisenhower Matrix** can help you categorise tasks based on urgency and importance**. Asana is a great web and mobile work management tool for those who need to help teams organise, track, and manage their work.** By focusing on high-priority tasks first, you can ensure that your most critical work is completed, leaving you more time to relax and explore.

### Create a structured routine

Establishing a daily routine can provide structure and help you stay disciplined. Start by setting specific working hours and sticking to them as closely as possible. Whether you're an early bird or a night owl, find a schedule that aligns with your natural productivity peaks. If you work with clients in America or Europe, keep this in mind and start work later in the day, after a leisurely morning of exploring, fitness, or relaxation. Incorporate regular breaks into your routine to prevent burnout and maintain focus. Having a consistent routine can also make it easier to plan leisure activities and social outings around your work commitments.

### Use productivity tools

Numerous productivity tools can help you manage your time more effectively. Project management apps like **Trello**, **Asana**, and **Monday.com** allow you to organise tasks, set deadlines, and track progress. Time-tracking apps like **Toggl** and **Clockify** can help you monitor how much time you spend on different activities, enabling you to identify areas where you can improve efficiency. Additionally, calendar apps like **Google Calendar** and **Microsoft Outlook** can help you schedule work tasks, meetings, and leisure activities in one place.

**Limit distractions**

Distractions can significantly impede your productivity, especially when working in an unfamiliar environment. Identify common distractions and take steps to minimise them. For example, if you find social media a major timewaster, use apps like **Freedom** or **StayFocusd** to block distracting websites during work hours. If noise is an issue, invest in noise-cancelling headphones or find a quiet workspace. Creating a dedicated workspace, even in your accommodation, can help you stay focused and separate work from leisure.

**Practice the Pomodoro Technique**

The Pomodoro Technique is a popular time management method that involves working in focused intervals, typically 25 minutes, followed by a short break. After completing four intervals, take a longer break. This technique can help you maintain concentration and prevent burnout. Use a timer or a Pomodoro app to keep track of your intervals and breaks. The regular breaks also allow you to stretch, hydrate, **and** recharge, making you more productive in the long run.

**Set boundaries**

Setting boundaries is essential for maintaining a healthy work-life balance. Communicate your working hours to clients, colleagues, and friends to ensure they respect your time. Avoid checking work emails or messages outside of your designated working hours. Setting boundaries also applies to yourself; resist the temptation to extend your workday unnecessarily and make a conscious effort to disconnect and enjoy your leisure time. Sign up for classes and courses that will force you to log off and make time for leisure activities.

**Delegate and outsource**

If your workload becomes overwhelming, consider delegating tasks or outsourcing certain responsibilities. Virtual assistants, freelance professionals, and specialised agencies can handle various tasks, from administrative work to content creation, allowing you to focus on more critical aspects of your job. **Fiverr, Freelancer, and Upwork are all fantastic platforms that connect freelancers to people or businesses looking to hire across a wide range of disciplines.** Do not work harder; work smarter. Delegating not only lightens your workload but also frees up time for leisure and exploration.

Scheduling work around travel and leisure

Balancing work with travel and leisure requires careful planning and flexibility. Here are some tips to help you schedule your work effectively while making the most of your time in Thailand.

**Plan ahead**

When planning your travels, consider your work commitments and deadlines. Use a calendar to map out your travel itinerary alongside your work schedule. Identify periods of high workload and plan your travel around them, ensuring ample time to complete your tasks. Conversely, during quieter work periods, take advantage of the opportunity to explore new destinations and engage in leisure activities.

**Allocate specific days for work and play**

One effective strategy is to allocate specific days or blocks of time for work and leisure. For example, you might dedicate weekdays to work and reserve weekends for travel and relaxation. Alternatively, you could work intensively for a few days and then take a day off to explore. By creating a clear distinction between work and play, you can fully immerse yourself in each activity without feeling guilty or distracted.

**Embrace flexibility**

While having a structured routine is important, it's equally crucial to embrace flexibility. Travel plans can change unexpectedly, and you may encounter opportunities for spontaneous adventures. Allow yourself the freedom to adjust your schedule when needed. If you find a particularly inspiring location, consider extending your stay and adjusting your work hours accordingly. Flexibility lets you maximise your experiences without compromising your professional responsibilities.

**Choose accommodation wisely**

Select accommodation that supports both work and leisure. Look for places with reliable internet, comfortable workspaces, and proximity to attractions and amenities. Co-living spaces and serviced apartments often provide ideal environments for digital nomads, offering both professional facilities and social opportunities. By choosing accommodation that meets your needs, you can seamlessly transition between work and relaxation.

**Take advantage of local experiences**

One of the joys of being a digital nomad in Thailand is the opportunity to immerse yourself in the local culture. Schedule time to participate in local experiences, such as cooking classes, cultural tours, fitness camps, games nights, and outdoor activities. These experiences not only enrich your stay but also provide a refreshing break from work. By integrating local activities into your schedule, you can create a fulfilling and well-rounded lifestyle.

**Set realistic expectations**

Be realistic about what you can achieve in a day. Avoid overloading your schedule with too many work tasks or leisure activities. Prioritise quality over quantity and permit yourself to enjoy downtime without feeling guilty. Setting realistic expectations ensures you can maintain a healthy balance and avoid burnout.

**Reflect, adjust, and stay organised**

Organisation is key to balancing work and play effectively. Keeping track of your tasks, deadlines, and travel plans can make a world of difference. Here are some tech tools and apps that can help you stay organised:

- **Trello**: A versatile project management tool that uses boards, lists, and cards to help you organise tasks and projects. Perfect for visual thinkers.
- **Todoist**: A powerful to-do list app that helps you keep track of your tasks and deadlines. It's great for setting reminders and prioritising your work.
- **Google Calendar**: An essential tool for scheduling and managing your time. You can set up multiple calendars, share them with others, and even integrate them with other apps.
- **Evernote**: An all-in-one note-taking app that allows you to capture ideas, create to-do lists, and save important information. It's perfect for keeping everything in one place.
- **Notion**: A comprehensive tool that combines notes, tasks, databases, and calendars. It's highly customisable and great for managing both work and personal projects.
- **Asana**: A robust project management app designed for teams but equally useful for individual use. It helps you track work, organise tasks, and set deadlines.

- **Expensify**: To manage your finances, Expensify helps you track expenses, scan receipts, and generate expense reports. It's a must-have for staying on top of your budget.

Regularly reflect on your time management practices and adjust as needed. Identify what's working well and what could be improved. Flexibility is key, as your routine may need to change based on your location, workload, travel plans, and personal preferences. Staying organised helps you stay on top of your responsibilities and frees up mental space to enjoy your leisure time. With these tools, you can keep everything running smoothly and focus on making the most of your digital nomad adventure in Thailand.

### Conclusion

Balancing work and play as a digital nomad in Thailand is both an art and a science. By implementing effective time management strategies and carefully scheduling your work around travel and leisure, you can enjoy the best of both worlds. Embrace the flexibility and freedom of the digital nomad lifestyle and make the most of your time in this beautiful and vibrant country. With some planning and a mindful approach, you can achieve a harmonious balance that maximises productivity and enriches your overall experience.

## NETWORKING & BUSINESS CULTURE

One of the key elements to thriving as a digital nomad in Thailand is understanding and integrating into the local business culture. Building strong professional relationships and grasping the nuances of Thai business etiquette can significantly enhance your experience and open doors to new opportunities. This section will guide you through making local connections and navigating the intricacies of Thai business culture, helping you create meaningful relationships and succeed in your work endeavours.

### Fitting In

Networking is essential for any professional, and as a digital nomad, building a robust network can provide support, collaboration opportunities, and valuable insights into the local market. Here are some strategies to help you make local connections.

## Making local connections

#### Attend networking events and meetups

Thailand boasts a vibrant community of entrepreneurs, freelancers, and digital nomads. Cities like Bangkok, Chiang Mai, and Phuket regularly host networking events, meetups, and conferences that cater to various industries. Websites like **Meetup.com**, **Eventbrite**, and local **Facebook Groups** can help you find events that align with your interests and professional goals. Attending these gatherings allows you to meet like-minded individuals, exchange ideas, and forge valuable connections.

#### Join co-working spaces

As discussed in previous sections, co-working spaces are not only excellent for productivity but also for networking. Many co-working spaces organise social events, workshops, and networking sessions, providing opportunities to interact with fellow professionals. Engage with the community by participating in these events and utilising shared spaces, such as communal areas and cafés, to create conversations and build relationships.

#### Leverage online communities

Online communities and forums are invaluable resources for networking and finding support. Platforms like **LinkedIn,**

**Nomad List**, and various **Facebook Groups** dedicated to digital nomads in Thailand offer a wealth of information and opportunities to connect with others. **Engage in discussions, ask for advice, and share your experiences** to build a presence within these communities. Virtual networking can be just as impactful as face-to-face interactions, especially in the digital age.

### Volunteer and participate in local initiatives

Volunteering for local organisations or participating in community initiatives can help you integrate into the local culture and connect with residents. Whether it's teaching English, joining environmental projects, or participating in cultural exchange programmes, volunteering allows you to contribute to the community while building meaningful relationships. These connections can lead to professional opportunities and a deeper understanding of Thai culture.

### Be a chamber member

Joining local Chambers of Commerce can provide unparalleled networking opportunities and support. These organisations host various exclusive events such as business luncheons, seminars, and social gatherings, offering platforms to connect with like-minded professionals and gain insights into the local business environment. As a member, you also gain access to industry leaders who share their knowledge on trends, challenges, and opportunities. Active participation can significantly increase your visibility, enhance your professional profile, and open doors to new collaborations and ventures. Some of the best ones are the **Australian Chamber of Commerce, Thailand (AustCham)**, **American Chamber of Commerce in Thailand (AmCham)**, **British Chamber of Commerce Thailand (BCCT)**, **German-Thai Chamber of Commerce (GTCC)**, and **Thai-French Chamber of Commerce (FTCC).**

**Collaborate with local businesses**

Collaborating with local businesses can provide mutual benefits and foster strong professional relationships. Whether partnering on projects, offering expertise, or seeking local services, collaboration can open doors to new opportunities. Approach local businesses that are genuinely interested in their work and explore ways to support each other. Building a reputation for being reliable and collaborative can enhance your standing within the local business community.

**Utilise networking apps**

Networking groups like **Shapr**, **Bumble Bizz**, **InterNations**, **Toastmasters**, and **Lunchclub** can help you connect with professionals in your area. These apps use algorithms to match you with individuals who share similar interests and goals, facilitating meaningful connections. Utilise these tools to expand your network and discover potential collaborators, mentors, or clients.

**Be open and approachable**

Building connections requires a proactive and approachable attitude. Be open to meeting new people, conversing, and sharing your experiences. Showing genuine interest in others and being willing to offer support or advice can go a long way in fostering strong relationships. Approach networking with a positive and open mindset, and you'll find that opportunities will naturally arise.

**Understanding Thai business etiquette**

Thai business culture is rooted in respect, harmony, and subtlety. Understanding the nuances of Thai etiquette can help you navigate professional interactions smoothly and build trust with local partners. One important concept to be aware of is 'saving face', which involves maintaining dignity and avoiding embarrassment for yourself and others. Here are some key aspects of Thai business etiquette to keep in mind:

**Saving face: A cultural cornerstone**

The concept of 'saving face' is deeply ingrained in many Asian cultures, including Thailand. Originating from China, the idea of face revolves around maintaining one's dignity, honour, and social standing. This cultural norm is prevalent across Asia and plays a significant role in personal and professional interactions.

In Thai business culture, saving face means avoiding actions or words that could cause embarrassment or loss of respect for oneself or others. This can include anything from public criticism to overt displays of anger. The goal is maintaining harmony and mutual respect, even in challenging situations. Understanding and practising this concept can help you build stronger, more respectful relationships with your Thai business partners.

**The wai greeting**

The traditional Thai greeting, known as the 'wai', involves placing your palms together in a prayer-like gesture and bowing slightly. The wai is a sign of respect commonly used in social and professional contexts. When meeting someone for the first time, especially an elder or someone of higher status, initiate the wai to show respect. While handshakes are becoming more common in business settings, the wai remains an important cultural practice.

**Respect for hierarchy**

Thai culture places a strong emphasis on hierarchy and respect for authority. In business settings, it's essential to recognise and respect the hierarchical structure. Address senior individuals with their appropriate titles and avoid questioning their authority in public. Showing deference to those in higher positions is a sign of respect and can help build positive relationships.

### Subtle communication

Thai communication tends to be indirect and subtle, emphasising maintaining harmony and avoiding confrontation. It's important to be mindful of your tone and body language, as overt expressions of anger or frustration are considered impolite. When discussing sensitive topics, approach the conversation with tact and sensitivity. Reading between the lines and paying attention to non-verbal cues can help you understand the true meaning behind words.

### Building personal relationships

In Thailand, building personal relationships is often a precursor to successful business dealings. Take the time to get to know your business partners personally and engage in social activities such as meals or cultural events. Establishing trust and rapport through personal connections can pave the way for smoother professional interactions. Be patient, as building strong relationships can take time, but the effort is well worth it.

### Punctuality and patience

Punctuality is valued in Thai business culture, so try to arrive on time for meetings and appointments. However, be prepared for some flexibility, as time schedules can be more relaxed compared to Western standards. Patience is also important, as decision-making processes may take longer due to the emphasis on consensus and careful consideration. Avoid rushing decisions or applying pressure, and demonstrate your patience and understanding.

### Dress code

Dressing appropriately is important in Thai business settings. Business attire is generally conservative, with men wearing suits and ties and women opting for modest dresses or blouses with skirts or trousers. When attending meetings, formal events, or visiting government offices (i.e. Immigration), dress professionally and neatly. In more casual settings, smart-casual attire is acceptable, but it's always better to err on the side of formality.

**Gift giving**

Gift-giving is a common practice in Thai business culture and is often used to show appreciation and respect. When presenting a gift, choose something modest and thoughtful, such as a small souvenir or a token of appreciation. Avoid extravagant or overly expensive gifts, which can be seen as inappropriate. Present the gift with both hands and a slight bow and express your gratitude.

**Dining etiquette**

Business meals are an important aspect of building relationships in Thailand. When dining with business partners, follow local customs and etiquette. Wait for the host to signal the meal's start and avoid discussing business matters until the host initiates the conversation. Use utensils appropriately and try a bit of everything offered to you as a sign of respect. Express gratitude for the meal and the company.

**Avoiding open conflict**

Thais generally dislike open conflict and prefer to resolve issues with a calm and composed demeanour. Open confrontations or aggressive behaviour can cause discomfort and lead to a loss of face for everyone involved. It's important to handle disagreements delicately and seek amicable solutions. A calm and respectful approach will be appreciated and can help maintain harmony in your professional relationships.

### The power of a smile

In Thailand, a smile goes a long way. Often referred to as the 'Land of Smiles', Thailand values the warmth and positivity that a simple smile can convey. Smiling can diffuse tension, show friendliness, and create a welcoming atmosphere. Whether you're greeting someone, expressing gratitude or navigating a challenging situation, a genuine smile can enhance your interactions and build goodwill.

Understanding and respecting the concept of saving face and offering a simple smile, along with other aspects of Thai business etiquette, will help you navigate professional interactions smoothly and build trust with local partners. This cultural awareness not only facilitates better communication but also strengthens your professional relationships, paving the way for success in your business endeavours in Thailand.

## Conclusion

Navigating the local business culture and building strong professional relationships are essential for success as a digital nomad in Thailand. By attending networking events, joining co-working spaces, leveraging online communities, and participating in local initiatives, you can create a robust network of connections. Understanding and respecting Thai business etiquette, such as the wai greeting, hierarchical structures, and subtle communication, will help you navigate professional interactions smoothly and build trust with local partners. Embrace the opportunity to integrate into the local culture, and you'll find that your experience as a digital nomad in Thailand is both enriching and rewarding.

### Stay tuned . . .

Now that we've explored how to build meaningful connections and navigate the business landscape in Thailand, it's time to turn our attention to another crucial aspect of digital nomad life:

financial management. In the next chapter, we'll delve into practical tips for managing your money while living in Thailand. From setting up a bank account and transferring money internationally to budgeting and protecting your finances, **Chapter 3: Finance & Budgeting** will provide the essential knowledge to ensure your financial well-being while enjoying your nomadic lifestyle.

Whether you're just starting your journey or are a seasoned nomad, understanding the financial intricacies of living abroad can make a significant difference in your overall experience. We'll cover everything you need to know to manage your funds effectively, avoid common pitfalls, and maximise your financial resources. Stay tuned as we equip you with the tools and information necessary to maintain financial stability and thrive in your new, adventurous life in Thailand.

# Chapter 3:
# Finance & Budgeting

### Managing Your Finances as a Digital Nomad

Welcome to Chapter 3, where we dive into the nitty-gritty of managing your money while living the dream in Thailand. Whether you're sipping a coconut on a pristine beach or navigating the bustling streets of Bangkok, keeping your finances in check is the key to a stress-free nomadic lifestyle. In this chapter, we'll cover everything from setting up a bank account and transferring money internationally to budgeting tips and protecting yourself from scams. Money matters don't have to be daunting; with the right information and planning, you'll be able to focus more on enjoying your adventure and less on worrying about your wallet. So, grab a notepad (or open your favourite finance app) and let's start making your financial life in Thailand as smooth and enjoyable as possible.

## BANKING & MONEY TRANSFER OPTIONS

Navigating the financial landscape as a digital nomad in Thailand can seem overwhelming at first, but understanding the basics of banking and money transfer options is crucial for a smooth experience. In this section, we'll guide you through setting up a bank account, choosing the right bank, and exploring various money transfer methods. From traditional bank transfers to innovative digital services and cryptocurrencies, we'll help you find the most cost-effective and reliable ways to manage your finances. By mastering these essentials, you can focus more on enjoying your adventures and less on financial worries. Let's dive into Thai Banking 101.

## Setting up a Bank Account in Thailand

Setting up a bank account in Thailand is one of the first practical steps you'll want to take as a digital nomad. It makes managing your finances much easier, from paying rent and utility bills to withdrawing cash without hefty international fees. Plus, the most popular local payment methods – QR code and bank transfer – won't work unless you have a local bank account. Thailand has become a modern, largely cashless society, so having access to local banking will save you a lot of hassle.

### Choosing the right bank

Thailand has a robust banking system with many reputable banks to choose from. Some of the most popular ones among expats include **Bangkok Bank (BBL)**, **Kasikorn Bank (KBank)**, **Siam Commercial Bank (SCB)**, and **Krungthai Bank (KTB)**. Each has its own perks and special offers, so it's worth doing a bit of research to see which one suits your needs best.

For instance, Bangkok Bank has a wide network of ATMs and branches, making it very convenient. Kasikorn Bank is known for its excellent online banking services, which can be a lifesaver when you're managing your finances on the go.

### Types of accounts, fees, and access

Typically, you'll be looking to open a savings account. These accounts often come with an ATM or debit card and online banking facilities, making it easy to manage your funds. Some banks also offer accounts specifically designed for foreigners, which might include additional perks like lower fees for international transfers.

- **ATM fees**: Be aware that while using your bank's ATMs is usually free, withdrawing cash from another bank's ATM may incur a fee ranging from **THB 20 to THB 50 per transaction**. Additionally, international ATM withdrawals can be much higher, often around **THB 200 to THB 400 per transaction**. There is also an inter-province ATM fee of **THB 20 to THB 40** for most banks.

- **Annual account fees**: Most banks charge a small annual fee for maintaining your account, typically ranging from **THB 200 to THB 500**. This fee usually covers the cost of your ATM or debit card and basic account maintenance.
- **Withdrawal limits**: ATMs in Thailand typically have a standard withdrawal limit of **THB 20,000 per transaction**. However, this limit can vary based on your own bank's policies and the specific ATM you are using. If you need to withdraw more than this amount, you can do an over-the-counter cash advance or withdrawal at a banking branch, which allows you to access higher amounts than the ATM limits.
- **Online banking**: Make sure to set up online banking and download the bank's app for easy access to your account. Online banking allows you to manage your finances from anywhere, transfer money, pay bills, and check your balance with ease. It's a convenient way to keep track of your spending and ensure your finances are always in order.
- **Banking apps**: You'll definitely want to download the banking app onto your mobile phone for ease of payment. With these apps, you can easily scan QR codes and make online payments while out and about. Nowadays, Thailand is a cashless society, and a banking app will save you a lot of hassle. These apps often come with additional features such as budgeting tools, transaction alerts, and the ability to freeze your card

instantly if it's lost or stolen. Having the app also means you can transfer money to friends or pay for services without carrying cash.

**What you'll need**

To open a bank account, you'll generally need the following documents:

- A valid passport.
- A visa that allows you to stay in Thailand (some banks may require a long-term visa).
- Proof of address in Thailand (this could be a utility bill or a rental contract).
- A minimum deposit (this can vary depending on the bank, but it's usually around **THB 500 to THB 1,000**).

Some banks might have additional requirements, like a work permit or a letter of recommendation, but these are less common. It's always a good idea to call ahead or check the bank's website to confirm what you'll need.

**The process**

Once you've gathered your documents, head to your chosen bank branch. The process is usually straightforward, and the staff are often very helpful. Here's a step-by-step guide:

- **Visit the bank:** Go to the branch with your documents. It's best to visit in the morning when the bank is less busy (especially for a mall branch).
- **Fill out forms:** The bank staff will provide the necessary forms to complete.
- **Submit documents:** Hand over your documents for verification.
- **Initial deposit:** Make your initial deposit to activate the account.

- **Receive your card:** You'll be issued an ATM or debit card, which you can usually collect on the same day or a few days later.

Before we move on, let's just talk about three other money-related matters: tipping, cash, and credit cards.

## Other Money Matters

### Tipping

Tipping in Thailand is generally not obligatory, but it's appreciated as a goodwill gesture. For instance, leaving a small tip of **THB 20 to THB 100** at a massage parlour, nail salon, or hairdresser/barber can go a long way in showing appreciation for good service. Similarly, go ahead and tip a parking attendant, tuk-tuk driver, tour guide, or security guard for helping you out.

In restaurants, tipping is not expected, but it's always nice to round your bill up or leave a small tip, e.g. **THB 20**. High-end restaurants and establishments might include a 10% service charge in the bill, so check. If service charges are included, additional tipping is unnecessary but still appreciated if you received exceptional service. For taxis, rounding up the fare to the nearest **THB 10 or 20** is a common practice. For hotel staff, such as bellboys and housekeeping, a tip of **THB 20 to THB** 50 per service is customary.

### Cash

As stated earlier, Thailand is a modern country which has fully embraced online/mobile banking. You'll only need cash for local markets, some restaurants/cafes, and smaller establishments. However, it's always a good idea to have some cash on hand for emergencies or for places that don't accept cards. Local markets, street food vendors, and small family-run businesses typically prefer cash payments.

While larger stores, shopping malls, and chain restaurants usually accept cards, some smaller establishments may not. Additionally, having cash is useful for public transport options like tuk-tuks and songthaews, which usually do not accept card payments. It's also handy for making small purchases, such as buying bottled water or snacks from convenience stores, and for any incidental expenses during your travels.

Carrying a mix of small and larger denominations can be helpful, as some vendors may not have change for larger bills. Always keep an emergency stash of cash in a secure place, separate from your wallet, in case you lose your primary source of funds.

**International credit cards**

Many smaller businesses will shy away from accepting international credit cards, and if they do, they might charge you an extra 3-5% to use them. They may also impose a minimum spend, e.g. **THB 1,500**. Credit cards are more commonly accepted in larger hotels, upscale restaurants, and major retail chains. However, relying solely on international credit cards can be inconvenient and costly due to these additional charges and minimum spending requirements.

**Local credit cards**

Getting a local credit card as a foreigner is not that easy; you will need a work permit and additional documentation. Here's a more detailed look at what you'll need:

- **Work permit**: A valid work permit is typically required. This demonstrates stable employment and legal permission to work in Thailand.
- **Proof of income**: You'll need to provide proof of income, such as salary slips or a letter from your employer. Some banks may require a minimum income threshold to be met.

- **Thai bank account**: You must have a Thai bank account with a history of transactions showing regular deposits and stable financial activity.
- **Residential address**: Proof of a permanent residential address in Thailand, such as a rental agreement or utility bill, is usually required.
- **Identification**: A valid passport with a visa that allows you to stay in Thailand is necessary. Some banks may also require additional identification documents.

The application process can be lengthy, and approval is not guaranteed. It's often simpler to rely on a debit card from your local bank account for everyday expenses. Debit cards linked to your Thai bank account are widely accepted. They can be used for purchases, online payments, and cash withdrawals without the additional fees associated with international credit cards.

Many Thai banks also offer debit cards with added benefits, such as rewards programs or cashback on certain purchases. These can be a convenient and cost-effective alternative to a credit card, providing ease of electronic payments while avoiding the complexities of obtaining a credit card.

By taking these steps and being aware of these tips, you'll find that managing your finances in Thailand has become much simpler and more efficient. Happy banking!

## Best Ways to Transfer Money Internationally

One of the challenges of being a digital nomad is finding cost-effective and reliable ways to transfer money internationally. Whether you're getting paid by overseas clients or sending money back home, you'll want to avoid hefty fees and poor exchange rates.

**Bank transfers**

While traditional bank transfers are a tried-and-true method, they're often not the most cost-effective. Thai banks typically charge a fee for international transfers, and the recipient bank might also levy a fee. Additionally, the exchange rates offered by banks are usually not the best.

However, bank transfers can be a good option if you're dealing with large sums of money (e.g. up-front rent and deposit) and need the security that banks provide. Make sure to compare the fees and exchange rates different banks offer before proceeding.

**Digital transfer services**

Digital (online and app) money transfer services like **Wise** (formerly TransferWise), **PayPal, Skrill,** and **Revolut** have become increasingly popular among digital nomads for their ease of use and competitive rates.

- **Wise**: Known for its transparent fee structure and real exchange rate, Wise is a favourite among expats. It's easy to set up an account, and transfers are usually quick and straightforward.
- **PayPal**: While convenient and widely accepted, PayPal's fees can add up, especially for larger transfers. However, it's a great option for smaller amounts and for receiving payments from clients.
- **Skrill**: Similar to PayPal, Skrill offers fast transfers but with varying fees depending on the destination and amount.
- **Revolut**: Revolut is an all-in-one finance app that is becoming popular among travellers. Available to people from Australia, Brazil, the European Economic Area (EEA), Japan, New Zealand, Singapore, Switzerland, the United Kingdom, and the United States, Revolut offers excellent exchange rates and low fees for international transfers. The app also provides features like budgeting

tools, cryptocurrency trading, and insurance options, making it a versatile choice for digital nomads.

- **Revolut Premium**: Here's a little secret... Revolut Premium offers some fantastic perks for frequent travellers. With a paid plan, you get priority customer support, everyday protection insurance, lounge access, and an exclusive card. Plus, you can enjoy higher limits on fee-free ATM withdrawals, making it easier to access your funds without extra costs. Premium users also get access to a range of travel insurance options, including medical and dental coverage, which can be a lifesaver while on the move.

**Cryptocurrencies**

For the tech-savvy, cryptocurrencies like Bitcoin and Ethereum offer an alternative way to transfer money internationally. Transactions are usually fast and can be cheaper, especially for large sums. One of the major advantages is that cryptocurrency transactions bypass traditional banking fees and exchange rates, making them a cost-effective option. Additionally, cryptocurrencies can be very secure, thanks to blockchain technology, which ensures that transactions are transparent and tamper-proof. However, the value of cryptocurrencies can be volatile, with prices fluctuating significantly over short periods. This volatility can be a double-edged sword: while it offers the potential for substantial gains, it also poses a risk of significant losses. It's essential to stay informed and understand the market dynamics before diving into crypto transfers.

Not all merchants or services accept cryptocurrencies, so it's not a substitute for traditional banking methods. However, the number of businesses accepting crypto is growing, and you can find dedicated platforms and exchanges in Thailand where you can convert your crypto to Thai Baht (THB) if needed. Always use reputable exchanges and wallets to ensure the security of your funds.

When using cryptocurrencies, it's also crucial to keep track of any tax obligations in your home country and Thailand, as crypto transactions may be subject to capital gains tax or other regulations.

**Remittance services**

Services like **Western Union**, **MoneyGram**, **WorldRemit**, and **xe.com** are also options, especially if you need to send money quickly. They're widely available and can be a lifeline in emergencies. However, their fees can be high, and the exchange rates are often less favourable.

Tips for cost-effective transfers:

- **Compare rates**: Use comparison websites to check the fees and exchange rates offered by different services.
- **Transfer in bulk**: If possible, transfer larger amounts less frequently to save on fees.
- **Stay informed**: Exchange rates fluctuate, so keep an eye on them and transfer money when the rates are favourable.
- **Use local services**: Some local Thai services offer competitive rates for international transfers. Ask fellow expats for recommendations.

## Conclusion

Managing your finances as a digital nomad in Thailand might seem daunting at first, but with the right strategies and tools, it can be quite straightforward. Setting up a bank account will make your day-to-day transactions smoother, and knowing the best ways to transfer money internationally will save you time and money. Remember, the goal is to spend less time worrying about your finances and more time enjoying your adventure in the Land of Smiles. Whether you're lounging on a beach in Phuket or exploring the vibrant markets of Chiang Mai, having a good handle on your money matters will make your experience all the more enjoyable. Happy travels and happy banking!

## COST OF LIVING INFO & TIPS

Understanding the cost of living in Thailand is crucial for budgeting and planning your stay. Thailand offers an affordable lifestyle compared to many Western countries, but costs can vary significantly depending on your lifestyle, location, and personal preferences. Major cities like Bangkok and tourist hotspots like Phuket tend to be more expensive, while smaller towns and rural areas offer a lower cost of living. Additionally, your choices in housing, dining, transportation, and entertainment can dramatically influence your monthly expenses. Whether you prefer a luxurious lifestyle or a more frugal approach, having a clear understanding of these costs will help you make informed financial decisions. Here's a detailed breakdown of typical expenses to help you plan your budget effectively.

### Typical Monthly Expenses

#### Rent

Renting accommodation in Thailand can vary widely depending on the type of property and its location. Here's a general overview:

#### Apartments

More budget-friendly, ranging from **THB 10,000 to THB 30,000 per month**. Apartments are a good option for budget-conscious nomads. They may not offer as many amenities as condos, but they are still comfortable and convenient. Remember that lower-priced apartments may not include furnishings, so you might need to budget for some basic furniture and appliances.

**Condominiums**

Prices can range from **THB 20,000 to THB 50,000 per month**, depending on the location and amenities. Condos often come with facilities such as swimming pools, gyms, and security. In prime locations like Bangkok's central business district, prices can be higher, while in smaller cities or suburban areas, you may find more affordable options.

**Houses**

Typically, it ranges from **THB 30,000 to THB 100,000 per month**, depending on size and location. Houses offer more space and privacy, making them ideal for families or those who prefer a quieter lifestyle. Suburban areas tend to be more affordable, while beachfront, centrally located, and pool villas can be more expensive.

**Townhouses**

Townhouses typically range from **THB 15,000 to THB 40,000 per month**. They provide a balanced option between apartments and full-sized houses, offering more space than an apartment but at a generally lower cost than a standalone house. Townhouses are ideal for small families or individuals seeking a bit more room without the higher expenses associated with larger homes. They are commonly found in suburban areas and offer a blend of affordability and comfort.

**Utilities**

Utilities in Thailand are generally affordable, but costs can add up depending on your usage and the type of accommodation you have. Here's what you can expect:

**Electricity**

Expect to pay around **THB 1,500 to THB 3,000 per month**, depending on usage. However, remember that while the government rate is THB 5 per kWh, many Airbnbs, condos, and apartments will charge **THB 7 or 8 per kWh**. Check your contract before signing; sometimes, this increase can be negotiated, but more often than not, it can't, and you just have to pay the surcharge. Air conditioning and a private pool can significantly increase your electricity bill, especially during the hot season. Also, keep in mind that 'off-peak' hours between 9.30pm and 8am have cheaper electricity.

**Water**

Usually inexpensive, around **THB 300 to THB 800 per month**. Water is typically billed based on usage, and costs are generally low. Some accommodations may include water in the rent, so it's worth checking your lease agreement. However, if you rent a property with a pool and have to run a water pump, that will hike up your monthly bill. Other hidden expenses could be a water filter, automatic sprinklers, etc. Always factor in these potential additional costs when budgeting for utilities.

**Internet**

High-speed internet costs around THB 600 to THB 1,000 per month. Thailand has a good internet infrastructure, and you can find reliable service providers offering various packages. Some accommodations may include internet in the rent, so it's worth clarifying this before signing a lease. You can find detailed info about internet connectivity in Chapter 2: Work & **Productivity**.

**Groceries**

The cost of groceries in Thailand can vary depending on where you shop and the types of products you buy. Here's a breakdown:

**Local markets**

Fresh produce and local goods are very affordable. You can expect to spend around **THB 3,000 to THB 5,000 monthly** on groceries if you shop at local markets. These markets offer various fruits, vegetables, meat, and seafood at lower prices than supermarkets. Shopping at local markets also allows you to experience Thai culture and support local farmers.

**Supermarkets**

Imported goods and branded items can be more expensive. Monthly grocery bills at supermarkets like Tesco Lotus or Big C can range from **THB 5,000 to THB 10,000**. Supermarkets offer a more extensive selection of international products, which can be convenient if you're looking for specific items from home. However, be prepared to pay a premium for these imported goods.

**Dining out**

Thailand is renowned for its delicious and affordable food. Eating out is a big part of Thai culture, and you'll find a wide range of dining options to suit every budget:

**Street food**

Extremely affordable, with meals typically costing between **THB 30 to THB 60**. Street food is a staple in Thailand, offering a variety of dishes such as *pad thai, som tam* (papaya salad), and grilled meats. It's not only cheap but also a great way to experience authentic Thai cuisine.

**Restaurants**

Local restaurants are reasonably priced, with meals costing around **THB 100 to THB 300**. These restaurants offer a wide range of Thai dishes, often in a more comfortable setting than street food stalls. Western restaurants can be more expensive, ranging from **THB 300 to THB 1,000 per meal**. Dining at

Western restaurants can be a treat if you're craving familiar flavours, but it can also add significantly to your monthly expenses.

### Cafes and coffee shops

Cafes and coffee shops are popular among digital nomads for working and socialising. A cup of coffee typically costs between **THB 50 to THB 150**, depending on the type of coffee and the establishment. Many cafes also offer light meals and snacks, making them a convenient option for a quick bite.

### Transportation

Getting around in Thailand is relatively inexpensive, and there are various transportation options available:

### Public transport

In Bangkok, public transport is efficient and affordable. The **BTS Skytrain, BRT Rapid Transit Bus**, and **MRT Subway** are popular options, with fares ranging from **THB 15 to THB 60 per trip**. Monthly passes are available for frequent travellers. There is no mass transit in any other Thai city. In other cities, you must rely on local buses, tuk-tuks, and songthaews (covered pickup trucks).

### Taxis and ridesharing

Taxis are widely available and reasonably priced, with fares starting at **THB 35**. Ride-sharing services like **Grab, Bolt**, and **inDrive** (note: no **Uber** in Thailand) are also popular and offer a convenient and cost-effective way to get around. Prices vary depending on distance and demand.

### Motorbike rentals

Renting a motorbike is a popular option for getting around, especially in smaller towns and islands. A daily rental is usually around **THB 300 (plus fuel)**. However, note that you'll need to leave your passport or a set amount of money as 'collateral'. Otherwise, monthly rentals typically cost between **THB 2,000 to THB 5,000**, with additional costs for fuel and insurance. It's important to have a valid licence (in English) and wear a helmet for safety. Make sure to obey local laws: wear a helmet, don't drive on the footpath, and obey traffic lights. Police stops are common, particularly in the evening around popular tourist spots, and failure to comply can result in fines ranging from **THB 500 to THB 2,000**.

**Car rentals**

For those who prefer the flexibility of driving, renting a car can be a great option. Major international rental companies like **Hertz**, **Avis**, **Sixt**, **EazyCar**, and **Budget** operate in Thailand alongside local providers. Prices vary depending on the type of vehicle and rental duration, but you can expect to pay around **THB 800 to THB 2,500 per day**. Alternatively, you can also explore private rental options, which can be more affordable. For an extended contract, private rentals usually cost between **THB 8,000 to THB 15,000 per month**, depending on the car's age and condition.

**Driving license**

If you plan to stay in Thailand long-term, obtaining a local Thai driving license for a car, motorbike, or both is highly recommended. It not only ensures legal compliance but also offers convenience and potentially lower insurance premiums. However, if you're visiting or staying short-term, ensure you have an **international driving permit (IDP)** or your home country's driving license in English. If it is not in English, a certified translation from your embassy is necessary. Comprehensive insurance is crucial, as driving in Thailand can be an adventure. Be prepared for challenging local driving habits, always stay cautious, and avoid driving under the influence. To obtain a local Thai driving license, you'll need to do the following.

**Required paperwork**

- Your passport with a valid visa.
- A valid **IDP** or your home country's driving license. If it's not in English, you need a certified translation.
- A medical certificate from a Thai doctor.
- Two passport-sized photos.
- Proof of residence, which can be obtained from an immigration office or your embassy.

**The process**

- Booking an appointment online through the **Department of Land Transport (DLT)** website.
- On the first day, you must submit your documents and undergo a medical check-up.
- On the second day, attending a short training session and passing a written test.
- Finally, passing a practical driving test.

**Differences between car and motorbike licenses**

- Car and motorbike licenses are separate in Thailand. You can apply simultaneously but will need to undergo separate practical tests for each.
- The paperwork required is generally the same for either license, so if you're applying for both, you'll need duplicate sets of your documents.
- The written test is the same for both, but if you apply for both types of licenses, you might have to take two practical exams – one for the car and one for the motorbike.

**Note:** Having a local license can be handy in other ways, too; you can use it as proof of residency and as a valid ID, e.g. when boarding a domestic flight.

**Healthcare**

Healthcare in Thailand is known for its high quality and affordability. Here's what you can expect:

**Public hospitals**

Public hospitals offer affordable medical care, but waiting times can be long. It's advisable to have a basic understanding of Thai, as not all staff may speak English. These hospitals provide essential services and are equipped with modern facilities, but the experience can be less comfortable than in private hospitals. Public hospitals are ideal for routine check-ups and minor health issues if you're on a budget. The cost of treatment in public hospitals is significantly lower than in private ones, making them an accessible option for many. Consultations and minor treatments can cost between **THB 100 to THB 500.**

**Private hospitals**

Private hospitals provide high-quality care with shorter waiting times and English-speaking staff. Costs are higher than public hospitals but still affordable compared to Western countries. Health insurance is recommended to cover potential medical expenses. Bangkok has several internationally accredited hospitals, such as Bumrungrad International Hospital and Bangkok Hospital, which are known for their state-of-the-art facilities and specialist services. These hospitals cater to expats and medical tourists, offering comprehensive healthcare services, including advanced diagnostics, surgery, and specialised treatments. The comfort and convenience of private hospitals make them a preferred choice for many expats. Consultations can range from **THB 1,000 to THB**

**3,000**, while more complex treatments can cost upwards of **THB 10,000.**

### Annual health checks

Thailand offers excellent packages for annual health checks, catering to different genders, ages, and requirements. These comprehensive check-ups can cover everything from basic blood tests to advanced imaging and screenings. Many hospitals offer 'executive check-up' packages, which include a full range of tests and consultations with specialists. Prices can vary, but you can expect to pay between **THB 5,000 to THB 50,000** for a complete head-to-toe health check. These packages provide great value for money and are an excellent way to monitor your health regularly.

### Pharmacies

Pharmacies are widely available and offer a range of over-the-counter medications and prescription drugs. Prices are generally lower than in Western countries, and pharmacists can provide advice on minor health issues. Major pharmacy chains like Watsons and Boots have branches in most cities and towns, making it convenient to fill prescriptions and purchase health-related products. Many medications that require a prescription in Western countries can be bought over-the-counter in Thailand, but it's always wise to consult a pharmacist or a doctor before taking any medication. Prices for common medications can range from **THB 50 to THB 300.**

### Dental care

Thailand is also known for its high-quality dental care, which is both affordable and accessible. Dental clinics are widespread, and many offer services such as cleanings, fillings, crowns, and even cosmetic procedures like teeth whitening. Dental care in Thailand is often much cheaper than in Western countries, making it a popular destination for dental tourism. For example, a routine dental cleaning can cost between **THB 700 to THB 1,500**, while more complex procedures like dental implants are also competitively priced, ranging from **THB 50,000 to THB 100,000**.

**Eye care**

Eye care in Thailand is of high quality and affordable, with numerous clinics and hospitals offering comprehensive eye care services. From routine eye exams to advanced treatments like LASIK surgery and cataract removal, you can find various services at competitive prices. Many eye care centres have English-speaking staff and use the latest technology. A basic eye exam can cost between **THB 500 to THB 1,500**, while LASIK surgery typically ranges from **THB 30,000 to THB 60,000** for both eyes. Cataract surgery can cost between **THB 20,000 to THB 40,000 per eye**. It's advisable to seek recommendations and read reviews to find a reputable eye care provider.

**Health insurance**

While healthcare in Thailand is affordable, having health insurance is highly recommended to cover any unexpected medical expenses. There are various insurance plans available that cater specifically to expats. These plans can cover a range of services, from routine check-ups to emergency medical treatments and hospital stays. Some popular health insurance providers for expats in Thailand include **Bupa**, **Aetna**, and **Cigna**. It's important to compare different plans and choose one that best suits your needs and budget. Premiums can range from **THB 20,000 to THB 100,000 per year**, depending on coverage and age.

**Mental health services**

Mental health services are also available in Thailand, with various private clinics and hospitals offering counselling and psychiatric care. Many of these services are provided by English-speaking professionals, making it easier for expats to seek help. Mental health support can be accessed through individual therapy sessions, support groups, and even online counselling services, which can be particularly beneficial for those dealing with the stresses of living abroad. Individual therapy sessions can cost between THB 1,500 and THB 3,000.

### Specialised treatments and medical tourism

Thailand is a popular destination for medical tourism, offering specialised treatments at a fraction of the cost compared to Western countries. Procedures such as cosmetic surgery, fertility treatments, and orthopaedic surgeries are highly sought after. Medical tourists can expect high standards of care and advanced medical technology. Many hospitals in Thailand have international patient departments that assist with travel arrangements, accommodation, and aftercare, ensuring a smooth and stress-free experience. Cosmetic surgery can range from **THB 50,000 to THB 300,000**, while fertility treatments can cost between **THB 100,000 to THB 300,000.**

### Vaccinations and preventative care

Before moving to or travelling around Thailand, it's advisable to ensure you are up-to-date with routine vaccinations such as measles, mumps, and rubella (MMR), diphtheria-tetanus-pertussis, and influenza. Depending on your travel plans, additional vaccinations for diseases like hepatitis A and B, typhoid, and rabies may be recommended. Preventative care is an essential part of staying healthy in Thailand, and local clinics and hospitals can provide these vaccinations. Vaccinations can cost between **THB 1,000 to THB 3,000 each.**

### Aged care

Thailand is increasingly becoming a popular destination for aged care, offering high-quality services at lower costs compared to Western countries. There are various aged care facilities, including assisted living, nursing homes, and retirement communities. These facilities provide comprehensive care, including medical services, physical therapy, and recreational activities. Costs can vary widely depending on the level of care and facilities provided, but generally, you can expect to pay between **THB 20,000 to THB 60,000 per month**. If bringing your parents over for retirement is something you're considering, know that affordable aged-care options are available.

**Emergency services**

In case of medical emergencies, Thailand has a network of emergency services that can be reached by dialling 1669. Ambulance services are available, but response times can vary depending on the location. It's crucial to have a contingency plan and know the nearest hospital or clinic in your area. Many private hospitals offer emergency services with quicker response times and higher levels of care. Ambulance services can cost between **THB 500 to THB 2,000**, depending on the distance and level of care required.

For a deeper understanding of hospitals and health insurance in particular, please head over to **Chapter 5: Health & Wellness** for the full details.

**Leisure and entertainment**

Thailand has a lot on offer, so you'll have no shortage of fun things to do during your time off. From keeping fit to shaking your booty and shopping up a storm, here's a rundown of top activities:

**Gyms and fitness centres**

Memberships typically cost between **THB 1,000 to THB 3,000 per month**. Many gyms offer modern facilities, fitness classes, and personal training services. Again, for full details, please go to **Chapter 5: Health & Wellness**.

### Outdoor activities

Thailand's natural landscapes provide opportunities for activities such as hiking, diving, and rock climbing. Costs vary depending on the activity and location, but many outdoor pursuits are affordable. In general, entrance fees to Thailand's national parks as a non-Thai range between **THB 200** and **THB 500 for adults** (half that for kids).

### Cultural activities

Visiting temples, museums, and cultural sites is a great way to learn about Thai culture. Entrance fees are usually low, ranging from **THB 20 to THB 200**. Some must-visit cultural sites include the Grand Palace in Bangkok, Wat Pho (the Temple of the Reclining Buddha), and the ancient city of Ayutthaya. Many temples offer free entry, but a donation is always appreciated. Participating in local festivals, such as Songkran (Thai New Year) and Loy Krathong (Festival of Lights), is another enriching way to immerse yourself in Thai culture.

### Nightlife

Thailand's nightlife scene is diverse, with options ranging from street bars to high-end nightclubs. A night out can cost anywhere from **THB 500 to THB 3,000**, depending on where you go and what you drink, and whether or not you take in a live performance. For example, **Phuket FantaSea** is **THB 2,000 per head**, **Muay That at Rajadamnern Stadium in Bangkok** is **THB 1,500-3,500 per head,** and **Pattaya Alcazar Cabaret Show in Pattaya** is **THB 900-1,800 per head**. Bangkok offers a vibrant nightlife with famous spots like Khao San Road for budget-friendly bars and Sukhumvit for upscale clubs. Chiang Mai has a more laid-back vibe with plenty of live music venues and night markets to explore. Pattaya is known for its beach clubs and Walking Street, offering a variety of entertainment options.

## Cinemas

Cinema ticket prices are generally affordable, ranging from THB 200 to THB 600, depending on the theatre and the type of seating (from seat to sofa). Some cinemas offer discounts on certain days of the week, so it's worth checking for special deals. Some offer an amazing ultimate package (THB 2,000) with a bed-like seat, pillows, blankets, food, and drink. Major cinema chains like SF Cinema and Major Cineplex provide a wide range of movies in both Thai and English, often with the latest releases from Hollywood. Many cinemas offer cheap films on Tuesdays with tickets costing only THB 100. Bangkok also has a slew of excellent independent cinemas that are worth checking out, especially if you like foreign films. Check out Bangkok Screening Room, Cinema Oasis, House Samyan, Lido Connect, and The Friese-Greene Club.

## Spas and wellness

Thailand is renowned for its spa and wellness industry. A traditional Thai massage can cost as little as **THB 200 to THB 500 for an hour**, while more luxurious spa experiences in high-end resorts can range from **THB 1,000 to THB 5,000** or more. Many spas offer packages that include massages, facials, and other treatments. Visiting a spa is not only a great way to relax but also to experience a significant aspect of Thai culture. Some of the best-value spa chains to look out for our **Let's Relax Spa**, **Oasis Spa**, **Asia Herb Massage & Spa**, **Health Land Massage**, and **Divana Spa**.

## Shopping

From bustling street markets to modern shopping malls, Thailand offers a variety of shopping experiences. Bangkok, in particular, is a shopping paradise. Some of the most popular destinations for shopaholics include **Chatuchak Weekend Market** (Friday-Sunday only), **MBK Center**, **Siam Square,**

**Siam Center, ICONSIAM**, **CentralWorld**, **Gaysorn Centre** (high-end), **The Platinum Shopping Mall** (wholesale), **Gateway Ekamai** (Japanese), **Siam Paragon**, **Fortune Town** (electronics), and **EM District (EmQuartier, EmSphere**, and **Emporium)**. Nationwide, you'll find **Central, Lotus's, BigC, The Mall, Terminal 21,** and **Robinson** malls in all major cities. Prices can vary widely depending on what you're shopping for. Street markets are great for affordable clothes, accessories, and souvenirs, while malls offer a mix of high-end brands and mid-range stores. Bargaining is common in markets, so don't hesitate to haggle for a better price.

**Festivals and events**

Thailand hosts numerous festivals and events throughout the year, which are both culturally enriching and entertaining. Some popular festivals include Songkran (Thai New Year) in April, Loy Krathong (Festival of Lights) in November, and the Chiang Mai Flower Festival in February. These events are usually free to attend, but you may want to budget for food, drinks, and any special activities associated with the festivities. For full details, please go to **Chapter 6: Cultural Insights & Learning Thai**.

## Conclusion

Navigating the cost of living in Thailand as a digital nomad can be exciting and manageable with the right information. From affordable rent options and efficient public transport to delicious street food and numerous leisure activities, Thailand offers a vibrant lifestyle that won't break the bank. By understanding typical expenses and planning your budget accordingly, you can enjoy all that this beautiful country has to offer without financial stress. Whether you're working from a bustling café in Chiang Mai or exploring the serene islands of the Andaman Sea, Thailand provides an enriching and affordable experience that makes every baht count. Happy budgeting and enjoy your stay in the Land of Smiles!

# TAX CONSIDERATIONS & OBLIGATIONS

As a digital nomad in Thailand, it's crucial to understand your tax obligations to avoid any legal complications and manage your finances effectively. Navigating the tax landscape can be complex, but being informed will help you stay compliant and make the most of your income. This section will cover essential aspects such as residency status, tax filing requirements, and potential tax treaties between Thailand and your home country. By understanding these key points, you can ensure you meet your tax responsibilities while optimising your financial situation. Let's delve into the details of your tax considerations in Thailand.

## Understanding Your Tax Responsibilities

### Residency status and tax liability

To determine your tax obligations in Thailand, it's important to understand your residency status and how it impacts your tax liability. This section will explain the criteria for tax residency, the applicable income tax rates, and the role of double taxation agreements (DTAs) in preventing you from being taxed on the same income in multiple countries.

- **Tax residency:** In Thailand, you're considered a tax resident if you stay in the country for 180 days or more within a calendar year. Tax residents are required to pay tax on income earned both within and outside of Thailand, while non-residents are only taxed on income derived from Thai sources.
- **Income tax:** The personal income tax rate in Thailand ranges from 5% to 35%, depending on your income bracket. It's important to keep track of your income and expenses to accurately report your earnings.
- **Double taxation agreements (DTAs):** Thailand has DTAs with several countries to prevent double taxation.

These agreements can help you avoid being taxed on the same income in both Thailand and your home country. It's essential to check if your country has a DTA with Thailand and understand the agreement's specifics.

### Reporting and paying taxes

Fulfilling your tax obligations in Thailand involves several steps, from obtaining the necessary identification numbers to filing your returns accurately. This section will guide you through the process of reporting and paying taxes, ensuring that you remain compliant with Thai tax laws.

- **Tax identification number (TIN):** If you're required to pay taxes in Thailand, you must obtain a TIN from the Revenue Department. This number is necessary for filing tax returns and other tax-related activities.
- **Filing tax returns:** Tax returns in Thailand are typically due by the end of March for the previous calendar year. You can file your tax return online through the Revenue Department's e-filing system or in person at a local tax office.
- **Deductions and allowances:** Thailand offers various deductions and allowances that can reduce your taxable income, such as personal allowances, health insurance premiums, and charitable donations. Make sure to keep records of these expenses to claim them on your tax return.

### Penalties for non-compliance

Failure to comply with local Thai rules and regulations can result in serious consequences. If you do not adhere to tax responsibilities, you may face fines, penalties, and interest charges on unpaid taxes. In severe cases, non-compliance can lead to your visa being revoked or not extended, jeopardising your ability to stay in Thailand. It's crucial to stay informed and meet all legal requirements to avoid these potential issues. Penalties can range from small fines to substantial amounts depending on the severity of the non-compliance. Additionally, persistent non-compliance or fraudulent activities can result in legal actions, including prosecution, which may lead to imprisonment. It's also possible for non-compliant individuals to be blacklisted, preventing future entry into Thailand. Ignorance of the law is not a defence, so taking proactive steps to understand and fulfil your tax obligations is essential for maintaining your residency and peace of mind.

## Resources for Tax Advice

Navigating tax regulations can be complex, especially when dealing with multiple countries. Here are some resources to help you stay compliant:

### Professional tax advisors

**Local tax advisors**: Hiring a local tax advisor in Thailand can provide expert guidance on Thai tax laws and help you ensure compliance. They can assist with tax planning, filing returns, and understanding your responsibilities. Fees for local tax advisors can range from **THB 5,000 to THB 20,000**, depending on the complexity of your tax situation. Here are some top tax companies in Thailand to check out:

- Expat Tax Thailand
- Siam Legal
- Thai Accounting Network

**International tax advisors**: If you have income from multiple countries, an international tax advisor can help you navigate the complexities of cross-border taxation and ensure you take advantage of any relevant DTAs. Expect to pay between **USD 500 to USD 2,000** for international tax advisory services. Here are some recommended companies:

- Bright!Tax
- Greenback Expat Tax Services
- Tax Samaritan

**Specialised expat advisors**: Some tax advisors specialise specifically in expat taxation issues. These professionals can offer tailored advice on managing taxes across multiple jurisdictions and help you understand your unique tax obligations as a digital nomad. Fees can range from **USD 300 to USD 1,500**, depending on the level of service required. Check out the below sites for more info:

- Double M
- Expat Tax Online
- Forvis Mazars Thailand

**Online resources**

- **Thai Revenue Department website**: The official website of the Thai Revenue Department offers information on tax laws, filing requirements, and online services. It's a valuable resource for understanding your tax responsibilities in Thailand.
- **Expat tax forums and communities**: Online forums and communities, such as **Reddit's r/tax** and expat-focused websites, can provide insights and advice from fellow digital nomads with experience with Thai taxes. Be sure to verify any information you find with a professional advisor.
- **Tax software**: There are various tax software solutions designed to help expats manage their tax filings. These platforms often provide step-by-step guidance and can help you ensure that you are compliant with both Thai and international tax laws. Some popular options include **TaxAct**, **TurboTax**, and **H&R Block**, with costs

ranging from **USD 50 to USD 200**, depending on the complexity of your tax situation.

**Educational resources**

- **Webinars and workshops**: Many organisations and tax advisors offer webinars and workshops on tax-related topics for expats. These can be a great way to learn about your responsibilities and ask questions directly to experts. Costs for webinars and workshops can range from **free to around USD 100**.
- **Books and guides**: There are several books and guides available that cover international taxation for digital nomads. These can provide a comprehensive overview of tax considerations and help you develop a solid understanding of your responsibilities. Books and guides can be purchased for between **USD 20 to USD 50**.
- **Online courses**: Numerous online platforms offer courses on international taxation specifically tailored for expats and digital nomads. Websites like **Coursera, edX**, and **Udemy** provide comprehensive courses to help you understand complex tax issues. Prices for these courses range from **USD 20 to USD 200**, depending on the depth and length of the course.

## Conclusion

Understanding and managing your tax responsibilities as a digital nomad in Thailand is essential for maintaining financial stability and avoiding legal issues. By familiarising yourself with Thai tax laws, seeking professional advice, and utilising available resources, you can ensure compliance and focus on enjoying your time in the Land of Smiles. Remember, staying informed and proactive about your tax responsibilities will help you make the most of your digital nomad lifestyle while avoiding unexpected surprises.

## PROTECTING YOUR FINANCES

As a digital nomad in Thailand, safeguarding your finances is crucial to ensure a worry-free experience. While Thailand is generally a safe country, being vigilant about potential scams and fraud is essential to protect your hard-earned money. Scammers often target unsuspecting individuals, especially foreigners who may be unfamiliar with local customs and regulations. This section will equip you with the knowledge to recognise common scams, take preventive measures, and seek help if you fall victim to fraud. By being proactive and informed, you can enjoy your digital nomad lifestyle in Thailand with peace of mind.

### Avoiding Scams and Fraud

As a digital nomad in Thailand, it's essential to be vigilant and aware of various scams and frauds that can target unsuspecting individuals. While Thailand is generally a safe country, being cautious can save you from potential financial pitfalls.

#### Common scams to watch out for

Listed below are some of the most common daily scams you should be on the lookout for:

- **Taxi and tuk-tuk scams**: Some drivers may try to overcharge you or take you on longer routes to increase the fare. Always insist on using the meter in taxis and agree on a price before starting your journey in a tuk-tuk.
- **Gem scams**: Be wary of strangers who approach you with 'too good to be true' deals on gemstones. These scams typically involve overpriced or fake gems.
- **Fake tour operators**: Some unscrupulous individuals pose as tour operators offering cheap trips. Always book tours through reputable companies and read reviews beforehand.

- **ATM skimming**: Be cautious when using ATMs, especially in tourist areas. Skimming devices can capture your card details. Use ATMs inside banks or well-lit, secure locations.
- **Accommodation scams**: Fake listings on rental websites can lead to financial loss. Always verify the legitimacy of the property and the landlord before making any payments. Use trusted platforms and read reviews.
- **Visa and work permit scams**: Be cautious of individuals or agencies offering to expedite your visa or work permit for a fee. Always go through official channels and consult the Thai Immigration Bureau for accurate information.
- **Temple scams**: Some scammers may pose as monks or temple workers asking for donations. While it's common to make donations at temples, be cautious if you feel pressured or if the situation feels off. Verify with the official temple staff before making any contributions.
- **Jet ski scams**: In tourist areas, some operators may rent out jet skis and then claim you damaged the equipment upon return, demanding hefty fees for repairs. Always inspect the equipment thoroughly and take photos before and after use. Consider renting from reputable companies.
- **Bar and nightlife scams**: Some bars may inflate your bill or add items you didn't order. Always check your bill carefully before paying. Sometimes, individuals may invite you to a bar and then leave you with an unexpectedly high bill. Stick to well-known establishments and avoid going to unfamiliar places with strangers.
- **Timeshare and holiday club scams**: Be wary of presentations offering free gifts or meals in exchange for

attending a timeshare or holiday club sales pitch. These offers often involve high-pressure sales tactics and may not deliver the promised benefits. Research the company thoroughly before committing to anything.

**Business scams**

While doing business in Thailand can be rewarding, it comes with its own set of challenges and potential scams. Navigating the local business landscape requires vigilance and awareness of common pitfalls. Here are some examples of scams and issues you may encounter:

- **Landlord scams**: Some landlords might hike the rent without warning or refuse to return your deposit. Always have a written contract, and make sure all terms are clearly stated. It's also wise to use a reputable real estate agent.
- **Police 'protection money'**: Unfortunately, some businesses may encounter local police officers demanding 'protection money'. This is illegal, and you should report such incidents to higher authorities or seek legal advice.
- **Fake business partners**: Be cautious of individuals posing as potential business partners who may have ulterior motives. Make sure to thoroughly check their background and history. It's a good idea to consult local business advisors or legal experts to ensure everything is legitimate.
- **Licensing and permits**: Ensure you have all the necessary licenses and permits to operate your business legally. Scammers might offer to expedite this process for a fee, but it's best to go through official channels to avoid complications.
- **Legal representation scams**: Some individuals may pose as legal representatives or consultants, offering to

help you navigate Thai business laws for a fee. Always verify their credentials and seek recommendations from trusted sources.

- **Supplier and contractor fraud**: When dealing with suppliers or contractors, ensure clear agreements and verify their reliability. Some may provide substandard goods or services or disappear after receiving payment.
- **Inflated service fees**: Be wary of providers who charge exorbitant fees for basic services. Always get multiple quotes and compare prices to ensure you're not overcharged.
- **Unauthorised withdrawals**: Ensure that any financial transactions are transparent and documented. Some unscrupulous partners might make unauthorised withdrawals or mismanage funds.

**Tips to avoid scams**

- **Do your research**: Before making any financial transactions or bookings, research thoroughly. Read reviews, check credentials, and consult trusted sources. Websites like **TripAdvisor, Google Reviews**, and **expat forums** can be very helpful.
- **Trust your instincts**: If something feels off or too good to be true, it probably is. Trust your gut and avoid suspicious offers.
- **Use reputable services**: Whether it's booking accommodation, tours, or transport, stick to reputable and well-known providers. Look for companies with established reputations and verified customer feedback.
- **Be cautious with strangers**: While Thai people are generally friendly and hospitable, be cautious of strangers who approach you with unsolicited offers or deals.

- **Verify official documents**: Always verify the authenticity of official documents like visas, work permits, and business licences. Use official government websites or visit local offices to confirm their validity.
- **Keep personal information secure**: Be mindful of sharing personal and financial information. Scammers can use this data for identity theft or fraudulent activities.
- **Negotiate and confirm prices**: Always negotiate and confirm prices beforehand for services like taxis, tuk-tuks, and tours. This helps avoid any surprises or inflated charges later.
- **Be aware of your surroundings**: Especially in crowded areas or tourist spots, be mindful of your belongings and potential pickpockets or scammers.
- **Use secure payment methods**: Whenever possible, use secure payment methods such as credit cards or reputable online payment platforms. Avoid carrying large amounts of cash.
- **Stay updated on common scams**: Scammers often develop new tactics. Stay updated on common scams by following local news, expat communities, and travel advisories.

By following these tips, you can significantly reduce the risk of falling victim to scams and enjoy a safer, more enjoyable experience in Thailand.

## Keeping Your Financial Information Secure

Before we end this chapter, let's discuss why protecting your financial information is crucial, especially when you're constantly moving. Here are some strategies to keep your financial data secure:

**Secure online transactions**

- **Use strong passwords**: Create strong, unique passwords for your online banking and financial accounts. Avoid using easily guessable information like birthdays or common words. Consider using a password manager to generate and store complex passwords, ensuring each account has a unique password for added security.
- **Enable two-factor authentication (2FA)**: Whenever possible, enable 2FA for an extra layer of security. This requires a second verification form, such as a code sent to your phone or an authentication app, in addition to your password. This reduces the risk of unauthorised access even if your password is compromised.
- **Beware of phishing scams**: Phishing scams can trick you into revealing personal information. Be cautious of emails, messages, or websites requesting your financial details. Always verify the source before clicking on links or providing information. Look for signs of phishing, such as generic greetings, spelling errors, and suspicious URLs.

**Protect your devices**

- **Keep software updated**: Regularly update your devices' operating systems and software to protect against security vulnerabilities. Software updates often include patches for security flaws that hackers could exploit. Enable automatic updates where possible to ensure you're always protected.
- **Install security software**: Use antivirus and anti-malware software to protect your devices from threats. Ensure that your security software is up-to-date and performs regular scans. Consider using a comprehensive security suite with firewall protection and real-time threat detection.

- **Avoid public Wi-Fi for financial transactions**: Public Wi-Fi networks can be insecure, making it easier for hackers to intercept your data. Use a virtual private network (VPN) when accessing financial accounts or conducting transactions on public networks. A VPN encrypts your internet connection, making it more secure.

**Physical security measures**

- **Safeguard your cards**: Keep your bank and credit cards secure. Consider using RFID-blocking wallets to protect against electronic pickpocketing. Be mindful of where you store your cards and avoid leaving them unattended in public places.
- **Monitor your accounts**: Regularly check your bank and credit card statements for suspicious transactions. Report any discrepancies to your bank immediately. Early detection of unauthorized transactions can help prevent further fraud and minimise financial loss.
- **Limit cash-carrying**: While it's good to have some cash on hand, avoid carrying large amounts. Use cards or digital payment methods for most transactions. This reduces the risk of theft and makes it easier to track your spending.

**Using financial tools**

- **Revolut and other finance apps**: Apps like **Revolut** offer various security features, such as instant spending notifications, the ability to freeze and unfreeze cards, and secure online shopping. Take advantage of these tools to monitor and control your spending. These apps often provide real-time alerts, helping you detect any unusual activity immediately.

- **Bank alerts**: Set up alerts with your bank to receive notifications for large transactions or unusual activity. This can help you detect and respond to potential fraud quickly. Customise your alert settings to cover various types of transactions, ensuring comprehensive monitoring.
- **Virtual cards**: Many banking apps and financial services offer virtual cards for online transactions. These cards can be used for one-time purchases and then deactivated, providing an additional layer of security against online fraud. Virtual cards are particularly useful for protecting your main account details during online shopping.

## Conclusion

Protecting your finances as a digital nomad in Thailand involves being vigilant against scams and fraud and taking proactive steps to secure your financial information. By staying informed about common scams, using reputable services, and incorporating robust security measures for your online and physical transactions, you can enjoy your nomadic lifestyle with peace of mind. Remember, a little caution goes a long way in ensuring your financial well-being. Enjoy your adventures in Thailand, knowing that your finances are secure and well-protected.

**Stay tuned . . .**

As we wrap up our chapter on finance and budgeting, we hope you feel more confident managing your finances while living and working in Thailand. Get ready to move on to **Chapter 4: Visa & Legal Requirements**, where we'll dive deep into the intricacies of Thai bureaucracy and legalities. We'll explore various visa options, such as Tourist, Smart, and Education Visas, ensuring you have a comprehensive understanding of each. From the nitty-gritty of application processes to outlining the pros and cons, we'll cover it all. You'll also learn about essential processes like 90-day reporting, which is crucial for staying compliant with local laws, and securing work permits if you decide to take up employment opportunities.

In addition to detailed guides and practical tips, this chapter will also cover important deadlines and procedures to remember so you're never caught off guard. Whether you're extending your stay, dealing with legal considerations for work permits, or just navigating the Thai legal landscape, you'll gain the knowledge and confidence to handle it all with ease. Get ready to tackle visa requirements and stay compliant while living your dream in this vibrant country.

# Chapter 4: Visa & Legal Requirements

### Living the Dream: Navigating Thai Paperwork with a Smile

So, you're dreaming of sipping on a coconut while working from a beach in Thailand? But before you can dive into the turquoise waters or explore the bustling streets of Bangkok, you need to get your paperwork in order. Navigating the Thai visa system might seem daunting at first, but don't worry – we've got you covered. In this chapter, we'll explore the different types of visas suitable for digital nomads and weigh the pros and cons of each. We'll also provide tips on how to apply and extend your stay. Ready? Let's get started.

## VISA OPTIONS FOR DIGITAL NOMADS

When it comes to staying in Thailand, there are a few visa options that digital nomads commonly consider: the Tourist Visa, Smart Visa, Non-Immigrant Visa, Thai Elite Visa, Education Visa, and Guardian Visa. Each has its own advantages and potential drawbacks, so it's important to choose the one that fits your lifestyle and work needs best.

### Introduction to Thai Visa Types

Navigating the world of Thai visas can seem daunting at first, especially given the variety of options available. The key to a smooth transition into your new life as a digital nomad in Thailand is understanding the different types of visas and selecting the one that aligns with your personal and professional needs. In this section, we will provide an overview of the primary visa types suitable for digital nomads, touching upon their specific requirements and purposes.

**Top 6 visa categories**

1. **Tourist Visa:** The Tourist Visa is one of the most straightforward options for those looking to stay in Thailand for a short period. It typically allows stays of up to 60 days and can often be extended for an extra 30 days. This visa is ideal for individuals who want to explore the country while working remotely but aren't planning a long-term stay. However, it's important to note that the Tourist Visa does not permit you to engage in any form of employment within Thailand.
2. **Smart Visa:** Designed for entrepreneurs, investors, skilled professionals, and startups in specific industries such as tech and innovation, the Smart Visa allows you to work without a traditional work permit. It offers longer stays and fewer reporting requirements, making it an excellent choice for individuals involved in high-growth sectors.
3. **Non-Immigrant Visa:** The Non-Immigrant Visa is more suited for those intending to stay in Thailand for an extended period. The 'B Visa' is specifically for business purposes and is often required if you plan to obtain a work permit. The 'O Visa' is generally for those accompanying family members working in Thailand or retirees. This visa type offers more flexibility and can be extended, making it a popular choice for digital nomads who foresee a longer stay.
4. **Thai Elite Visa:** The Thai Elite Visa is a premium option that offers long-term residency for up to 20 years, depending on the package you choose. It comes with a hefty price tag but offers a range of VIP benefits, including expedited immigration procedures, exclusive services, and various discounts. This visa is ideal for those who want to avoid the hassle of frequent renewals and enjoy a more luxurious lifestyle.

5. **Education Visa:** If you are interested in combining your stay with some form of educational pursuits, such as learning Thai or taking other courses, the Education Visa might be the right option for you. This visa allows you to stay in the country for your course and can be renewed as long as you remain enrolled. It's a great way to immerse yourself in the culture while complying with visa regulations.
6. **Guardian Visa:** The Guardian Visa applies to individuals who are guardians of students enrolled in Thai educational institutions. If you have children studying in Thailand, this visa allows you to stay in the country to care for them. While not as commonly used by digital nomads, it can be a viable option for those with family commitments.

**Tips for smooth running**

Applying for a visa can be a bit of a bureaucratic adventure, but these tips will help ensure your journey is as smooth as possible.

- **Start early**: Begin gathering your documents and preparing your application well in advance. This gives you plenty of time to address any issues that might arise.
- **Double-check requirements**: Visa requirements can vary between different embassies and consulates. Always check the specific requirements for the location where you're applying.
- **Keep copies**: Make copies of all your documents before submitting them. This includes your application form, passport, photos, and any additional paperwork. It's also a good idea to keep digital copies saved securely.
- **Be honest**: Provide accurate and truthful information on your application form and during any interviews.

Misrepresentations can lead to visa denial or future travel issues.

- **Stay organised**: Keep all your documents neatly organised in a folder. This will make it easier to find what you need and present a professional image during your application process.
- **Follow instructions**: Carefully read and follow all instructions provided by the embassy or consulate. This includes guidelines for photo specifications, document submission, and fee payments.
- **Seek assistance if needed**: If you're unsure about any part of the application process, don't hesitate to reach out to the embassy or consulate for clarification. There are also visa agencies that can assist with the process for a fee.
- **Prepare for the interview**: If an interview is required, dress appropriately and be prepared to discuss your plans in Thailand. Bring all your documents and be ready to provide additional information if requested.
- **Stay patient**: Visa processing times can vary, and delays can happen. Plan your travel dates accordingly and avoid booking non-refundable tickets until you have your visa in hand.
- **Stay updated**: Visa regulations can change, so it's important to stay updated on the latest requirements and procedures. Check the embassy or consulate's website regularly for any updates.
- **For Education Visa applicants**: Ensure you maintain regular attendance at your educational institution. Immigration authorities can check your attendance records and quiz you on your studies to verify that you are genuinely studying.
- **For Guardian Visa applicants**: Be prepared to provide comprehensive documentation proving your relationship

to the child and their enrolment in school. Only one Guardian Visa is allowed per child, so plan accordingly if both parents wish to stay.

**Pricing information**

Whilst this might vary from country to country, here we give a rough estimate of how much it might cost to obtain each type of visa.

1. **Tourist Visa**: The cost for a Tourist Visa generally ranges from **THB 1,000 to THB 2,000** for a single-entry visa, depending on the country of application. A multiple-entry Tourist Visa can cost around **THB 5,000**.
2. **Smart Visa**: The application fee for a Smart Visa is typically around **THB 10,000**, but this can vary depending on the category and duration.
3. **Non-Immigrant Visa**: The cost for a Non-Immigrant Visa is generally around **THB 2,000 to THB 5,000**, depending on the type and duration. Additional fees may apply for work permits.
4. **Thai Elite Visa**: Thai Elite Visa packages start at around **THB 800,000** for a 5-year membership and can go up to **THB 2 million** for a 20-year membership.
5. **Education Visa**: The cost for an Education Visa is typically around **THB 5,000**, but this can vary depending on the educational institution and course duration.
6. **Guardian Visa**: The application fee for a Guardian Visa is generally around **THB 5,000**, depending on the country of application and the duration.

## VISAS IN DETAIL

Now that we know what visas are available, how to apply, and what the costs are, let's take a deep dive into the six main categories to learn more about each one in detail. This will help you make an informed decision and ensure you choose the visa that best fits your needs and circumstances.

## Tourist Visa

The Tourist Visa is usually the go-to option for many first-time visitors. It's relatively easy to obtain and allows you to stay in Thailand for a decent amount of time without much fuss.

**Validity**: Typically, a single-entry Tourist Visa allows for a **60-day stay**, which can be extended for an additional 30 days at a local immigration office. This gives you a total of **90 days** in Thailand.

- **Multiple-entry option**: If you plan on coming and going from Thailand, you can apply for a multiple-entry Tourist Visa, which allows you to enter and exit the country multiple times within a six-month period, with each stay lasting up to **60 days**.

**Pros**

- **Ease of application**: The application process is straightforward. You can apply at any Thai embassy or consulate outside of Thailand.
- **Flexibility**: Ideal for those who want to explore Thailand and neighbouring countries.
- **Cost-effective**: Generally cheaper than other visa types, especially for short stays.

**Cons**

**Short-term solution**: Not ideal for long-term stays. You'll need to do visa runs or leave the country and re-enter to reset your stay.

- **Work restrictions**: Officially, you're not allowed to work on a Tourist Visa, even if your work is online and for a company outside of Thailand.

- **Frequent renewals**: The need for frequent visa extensions or renewals can be cumbersome and time-consuming, particularly if you intend to stay in Thailand for an extended period.

Certain passport holders can enter Thailand under a visa exemption scheme and receive a **30-day visa waiver** on arrival, which can be extended for an additional 30 days at a local immigration office. For a **60-day Tourist Visa** obtained before arrival, the following countries are among those eligible: Australia, Canada, USA, UK, and most European Union countries. Please check the latest information from the Thai embassy or consulate in your country, as eligibility and regulations can change.

**Step-by-step guide to applying**

- **Choose your embassy or consulate**: Find the nearest Thai embassy or consulate where you will submit your application. This can often be done in your home country or any other country where you are legally residing.
- **Gather your documents**: Ensure you have all the necessary documents ready (detailed below).
- **Fill out the application form**: Complete the visa application form, which is usually available on the embassy or consulate's website.
- **Submit your application**: Submit your application in person or by mail, depending on the consulate's requirements.
- **Pay the visa fee**: The fee can vary depending on the embassy or consulate and whether you are applying for a single-entry or multiple-entry visa.
- **Wait for processing**: Processing times can vary, but it typically takes a few days to a week.

- **Collect your visa**: Once approved, collect your visa from the embassy or consulate. Or they'll send it back to you if you apply by mail.

**Necessary documents**

- **Passport**: A valid passport with at least six months remaining validity and at least one blank page.
- **Visa application form**: Available on the embassy or consulate's website.
- **Passport-sized photos**: Typically two photos, meeting the specific size requirements outlined by the embassy.
- **Proof of accommodation**: Hotel bookings or an invitation letter from a friend or family member in Thailand.
- **Proof of financial means**: Bank statements showing sufficient funds to cover your stay (usually around **THB 20,000** per person).
- **Flight itinerary**: A copy of your return or onward travel ticket.

## How to extend your stay

If you're in Thailand on a Tourist Visa, you can extend your stay by an additional 30 days. Here's how you can do it without breaking a sweat:

**Prepare your documents**

- **Passport**: Ensure your passport is valid for at least six months beyond your intended stay.
- **Visa extension application form (TM.7)**: Available at immigration offices or for download online.
- **Passport-sized photos**: Typically, two photos that meet the specifications outlined by the immigration office.

- **Proof of financial means**: Bank statements showing sufficient funds (generally around **THB 20,000** per person).
- **Proof of accommodation**: Hotel bookings or an invitation letter from a friend or family member in Thailand.
- **Application fee**: The extension fee is **THB 1,900**, payable in cash.

**Visit the Immigration Office**

- **Locate the nearest immigration office**: In major cities like Bangkok, Chiang Mai, and Phuket, there are dedicated visa extension units.
- Arrive early to avoid long queues. Immigration offices can get quite busy, especially during peak tourist seasons.

**Submit your application**

- Complete the **TM.7** form and attach your photos.
- Submit your application along with the required documents and fee.
- An immigration officer will review your application, and you may be asked a few questions about your stay.

**Wait for processing**

- Processing times can vary, but it usually takes a couple of hours. Some offices may ask you to return later in the day or the next day to collect your passport.

**Collect your passport**

- Once your extension is approved, collect your passport with the new visa stamp. Congratulations, you've just extended your stay by 30 days!

**Important deadlines and procedures**

**Extension deadline**: Apply for your extension at least one week before your current visa expires to allow sufficient processing time.

- **Overstay penalties**: Overstaying your visa is a serious offence. Penalties include a fine of **THB 500 per day**, up to a maximum of **THB 20,000**. Overstays of more than 90 days can result in being blacklisted from re-entering Thailand for up to one year or more.

## Smart Visa

The Smart Visa is designed for entrepreneurs, investors, skilled professionals, and startups in specific industries such as tech and innovation. It offers a streamlined process for those who meet the eligibility criteria and provides several benefits not available with traditional visas.

- **Validity:** The Smart Visa can be issued for 1 to 4 years, depending on the category and the applicant's qualifications. It also includes a grace period of up to 60 days after the end of employment or project.
- **Multiple-entry option:** The Smart Visa allows for multiple entries into Thailand, making it convenient for frequent travellers and those with business interests in the region.

**Pros**

- **Work permit exemption:** Smart Visa holders are exempt from the requirement to obtain a separate work permit, simplifying the process of legally working in Thailand.
- **Longer stays:** The visa offers a longer validity period compared to other visa types, reducing the need for frequent renewals.

- **Family inclusion:** Dependents of Smart Visa holders, such as spouses and children, are also granted permission to stay in Thailand.
- **Fewer reporting requirements:** Smart Visa holders must report their stay to immigration only once a year, as opposed to every 90 days.

**Cons**

- **Eligibility criteria:** The Smart Visa has stringent eligibility requirements, including specific qualifications, industry focus, and employment conditions. Not everyone will qualify.
- **Higher application complexity:** The application process for the Smart Visa can be complex and may require substantial documentation and verification.
- **Cost:** While the Smart Visa provides numerous benefits, the application fees and associated costs can be higher compared to other visa types.

**Step-by-step guide to applying**

- **Check your eligibility:** Ensure you meet the criteria for the Smart Visa, which includes specific qualifications, industry focus, and employment conditions. Detailed eligibility information can be found on the official Smart Visa website.
- **Gather your documents:** Ensure you have all the necessary documents, including proof of qualifications, employment, or investment (detailed below).
- **Apply through the Smart Visa Unit:** Submit your application online via the official Smart Visa website or at designated Smart Visa Units. The application form and detailed instructions are available online.

- **Pay the application fee:** The fee varies depending on the Smart Visa category and duration. Ensure you pay the correct fee as specified.
- **Wait for processing:** Processing times can vary, but it typically takes a few weeks. During this time, your documents will be verified, and your eligibility will be assessed.
- **Attend an interview (if required):** Some applicants may be required to attend an interview to further verify their qualifications and intentions. Be prepared to discuss your plans and provide additional documentation if requested.
- **Receive approval and collect your visa:** Once approved, you will receive your Smart Visa approval letter. You can then collect your Smart Visa from the designated Smart Visa Unit or process it at a Thai embassy or consulate.

**Necessary documents**

- **Passport**: A valid passport with at least six months remaining validity and at least one blank page. Ensure your passport is up-to-date and renew it if necessary before applying.
- **Visa application form**: Available on the official Smart Visa website. Download and complete the form accurately, providing all required information.
- **Passport-sized photos**: Typically three photos, meeting the specific size requirements outlined by the Smart Visa Unit. These can be obtained at most photo studios or passport photo services.
- **Proof of qualifications**: Documentation that verifies your qualifications, such as degrees, certificates, or professional licenses. Obtain these from your educational institutions or professional bodies.

- **Employment or investment proof**: Depending on your Smart Visa category, provide evidence of your employment, investment, or startup activities. This may include:
- **Employment contracts**: Provided by your employer.
- **Company registrations**: Obtain these from your business or company.
- **Proof of investment**: Bank statements or investment documents showing your financial involvement in a Thai business or startup.
- **Health insurance**: Proof of health insurance coverage that meets the requirements specified for the Smart Visa. You can obtain suitable health insurance from international or Thai insurance providers.
- **Letter of recommendation**: If applicable, a letter of recommendation from relevant government agencies or organisations supporting your application. This can be obtained from entities such as the Thailand Board of Investment (BOI) or industry-specific organisations.
- **Application fee**: Be prepared to pay the application fee, which varies depending on the Smart Visa category and duration. Ensure you have the necessary funds available for this payment.

## How to extend your stay

### Prepare your documents

- **Passport**: Ensure your passport is valid for the duration of your extended stay.
- **Smart Visa extension form**: Available at the Smart Visa Unit or for download online.
- **Passport-sized photos**: Typically, three photos that meet the specifications outlined by the Smart Visa Unit.
- **Proof of ongoing qualifications**: Documentation verifying that you continue to meet the Smart Visa

criteria, such as employment contracts, company registration, or proof of investment.

- **Health insurance**: Updated proof of health insurance coverage that meets the Smart Visa requirements.
- **Application fee**: The extension fee varies depending on the Smart Visa category and duration, payable in cash or through other specified methods.

**Contact Thailand Elite Member Services**

- **Reach out to member services**: Inform them of your intention to extend your stay. They will guide you through the process and arrange for any necessary paperwork.
- **Concierge services**: Thailand Elite provides concierge services to assist you with immigration procedures, including arranging appointments and accompanying you if needed.

**Visit the Immigration Office**

- Smart Visa holders can visit designated Smart Visa Units for extensions.
- Arrive early to avoid long queues, especially during peak periods.

**Submit your application**

- Attach your photos and all required documents.
- Along with the required documents and fee. An immigration officer will review your application, and you may be asked questions about your ongoing qualifications.

**Wait for processing**

- Typically expedited due to the VIP services, but it usually takes a few days. Some offices may ask you to return later in the day or the next day to collect your passport.

**Collect your passport**

- Once your extension is approved, collect your passport with the new visa stamp. Congratulations, you've successfully extended your stay in Thailand!

**Important deadlines and procedures**

- **Extension deadline**: Apply for your Smart Visa extension at least one month before your current visa expires to allow sufficient processing time. This helps ensure that there are no gaps in your legal stay and avoids any last-minute issues.
- **Overstay penalties**: Overstaying your Smart Visa is a serious offence and incurs the same penalties as other visa types. Penalties include a fine of **THB 500 per day**, up to a maximum of **THB 20,000**. Overstays of more than 90 days can result in being blacklisted from re-entering Thailand for up to one year or more. Additionally, overstaying can negatively impact your ability to renew the Smart Visa or apply for other visas in the future.
- **Ongoing qualifications**: Continuously ensure you meet the eligibility criteria for the Smart Visa category under which you applied. This may involve maintaining employment, investment, or business activities as initially declared. Failure to meet these ongoing requirements can affect your visa status and extension eligibility.

## Non-Immigrant Visa

For those looking to stay longer in Thailand, the Non-Immigrant Visa might be a better fit. There are several types of

Non-Immigrant Visas, but the **B Visa** (for business) and the **O Visa** (for other purposes) are most relevant to digital nomads.

- **Non-Immigrant B Visa**: This visa is typically for those who have a job offer from a Thai company or are conducting business in Thailand. It can be a bit tricky for digital nomads unless you have a local sponsor (e.g. a company based in Thailand).
- **Non-Immigrant O Visa**: This visa is often used by those who are married to a Thai national, volunteering, or retiring in Thailand. Digital nomads sometimes use this visa if they qualify under one of these categories.

**Pros**

- **Longer stays**: Non-Immigrant Visas can be extended for up to one year, making them ideal for those who want to stay in Thailand long-term.
- **Work permits**: With the right paperwork, you can obtain a work permit, making it legal to work in Thailand.
- **Flexibility for various purposes**: The O Visa caters to a range of situations, including family unification and retirement, providing more options for different personal circumstances.

**Cons**

**Complex application process**: Requires more documentation and sometimes a local sponsor.

- **Higher costs**: Generally more expensive than a Tourist Visa, especially when you factor in the costs of obtaining a work permit.
- **Stringent requirements:** Meeting the eligibility criteria can be challenging, and the process may involve multiple steps and approvals.

**Step-by-step guide to applying**

- **Determine your eligibility**: Ensure you meet the criteria for the specific Non-Immigrant Visa type you are applying for (B for business, O for other purposes).
- **Gather your documents**: Collect all necessary documents, including any additional paperwork specific to the visa type (detailed below).
- **Complete the application form**: Fill out the visa application form, available on the embassy or consulate's website.
- **Submit your application**: Submit your application in person at the Thai embassy or consulate. Some consulates may allow for postal applications, but this varies.
- **Pay the visa fee**: Fees can vary, and you may need to pay additional fees for a work permit if applicable.
- **Attend an interview**: Some embassies or consulates may require an interview. Be prepared to discuss your plans and provide additional documentation if requested.
- **Wait for processing**: Processing times can vary, often taking a few days to a couple of weeks.
- **Collect your visa**: Once approved, collect your visa from the embassy or consulate.

**Necessary documents**

- **Passport**: A valid passport with at least six months remaining validity and at least one blank page.
- **Visa application form**: Available on the embassy or consulate's website.
- **Passport-sized photos**: Typically two photos, meeting the specific size requirements outlined by the embassy.
- **Proof of accommodation**: Hotel bookings or an invitation letter from a friend or family member in Thailand.

- **Proof of financial means**: Bank statements showing sufficient funds to cover your stay.

**Additional documents for B Vis**

- **Employment contract**: A signed contract from a Thai company.
- **Company registration documents**: Provided by your employer.
- **Letter of invitation**: From the Thai company.

**Additional documents for O Visa**

- **Marriage certificate**: If applying based on marriage to a Thai national.
- **Retirement documents**: Proof of retirement income if applying for retirement purposes.

## How to extend your stay

**Prepare your documents**

- **Passport**: Ensure your passport is valid for at least six months beyond your intended stay.
- **Visa extension application form (TM.7)**: Available at immigration offices or for download online.
- **Passport-sized photos**: Typically, two photos that meet the specifications outlined by the immigration office.
- **Proof of financial means**: Bank statements showing sufficient funds as required by your visa type.
- **Proof of accommodation**: Lease agreement or proof of residence.
- **Additional documents**: Depending on your visa type, you might need additional documents such as a letter from your employer, marriage certificate, or proof of retirement income.
- **Application fee**: The extension fee is **THB 1,900**, payable in cash.

**Visit the Immigration Office**

- Locate the nearest immigration office that handles Non-Immigrant Visa extensions. In major cities, there are specific offices for this purpose.
- Arrive early to avoid long queues.

**Submit your application**

- Complete the **TM.7** form and attach your photos.
- Submit your application along with the required documents and fee.
- An immigration officer will review your application, and you may be asked a few questions.

**Wait for processing**

- Processing times can vary, but it usually takes a couple of hours. Some offices may ask you to return later in the day or the next day to collect your passport.

**Collect your passport**

- Once your extension is approved, collect your passport with the new visa stamp. You've successfully extended your stay!

**Important deadlines and procedures**

- **Extension deadline**: Apply for your extension at least one month before your current visa expires to allow sufficient processing time.
- **Overstay penalties**: Similar to the Tourist Visa, overstaying a Non-Immigrant Visa incurs a fine of **THB 500 per day**, up to a maximum of **THB 20,000**. Longer overstays can lead to being blacklisted.

## Thai Elite Visa

If you're looking for a hassle-free, long-term solution and don't mind spending a bit more, the Thai Elite Visa might be just what you need. The Thai Elite Visa is essentially a long-term residence visa designed for affluent individuals looking to stay in Thailand for extended periods without the usual visa headaches.

- **Validity**: The Thai Elite Visa is available in various packages, ranging from **5 to 20 years**.
- **Exclusive benefits**: Holders of the Thai Elite Visa enjoy a range of benefits, including expedited immigration procedures, VIP services at airports, and access to exclusive events and discounts.

**Pros**

- **Long-term stay**: Perfect for those planning to stay in Thailand for several years.
- **Convenience**: No need for visa runs or frequent renewals. The visa is multi-entry, allowing you to come and go as you please.
- **VIP treatment**: Access to a range of exclusive services and benefits.

**Cons**

- **Cost**: The Thai Elite Visa comes with a hefty price tag, making it an option only for those who can afford the luxury.
- **Work restrictions**: Like the Tourist Visa, the Thai Elite Visa doesn't automatically grant you a work permit. You must apply separately if you wish to work legally in Thailand.
- **High up-front investment**: The visa requires a significant initial financial commitment, which may not be practical for everyone considering a long-term stay.

**Step-by-step guide to applying**

- **Apply online**: Visit the Thailand Elite website and complete the online application form.

- **Submit required documents**: Upload scanned copies of your passport and other necessary documents.
- **Pay the membership fee**: The fee varies depending on the package you choose (5, 10, or 20-year options).
- **Wait for approval**: The approval process can take a few weeks, during which the Thailand Elite team will verify your application.
- **Receive your membership card**: Once approved, you will receive your Thailand Elite membership card and visa.
- **Activate your visa**: Upon arrival in Thailand, present your membership card at the immigration counter to activate your visa.

**Necessary documents**

- **Passport**: A valid passport with at least six months remaining validity and at least one blank page.
- **Visa application form**: Available on the Thailand Elite website.
- **Passport-sized photos**: Typically two photos, meeting the specific size requirements.
- **Proof of financial means**: Bank statements showing sufficient funds to cover the membership fee.
- **Membership application**: Completed online via the Thailand Elite website.

## How to extend your stay

### Contact Thailand Elite Member Services

- Reach out to Thailand Elite's member services to inform them of your intention to extend your stay.
- They will guide you through the process and arrange any necessary paperwork.

**Prepare your documents**

- **Passport**: Ensure your passport is valid for the duration of your stay.
- **Elite Membership Card**: Have your Thailand Elite membership card ready.

**Visit the Immigration Office**

- Thailand Elite provides concierge services to assist you with immigration procedures.
- They will arrange an appointment at the immigration office and accompany you if needed.

**Submit your application**

- With the assistance of Thailand Elite staff, submit your application and required documents.
- The application fee is typically covered under your Elite membership benefits.

**Wait for processing**

- Processing is usually expedited, thanks to the VIP services.
- You may be able to collect your passport with the new visa stamp on the same day.

**Important deadlines and procedures**

- **Renewal deadline**: Thailand Elite members typically receive reminders from member services regarding renewal dates. Ensure you start the renewal process well before your current visa expires.
- **Overstay penalties**: Overstaying a Thai Elite Visa incurs the same penalties as other visas. However, given the VIP nature of the Elite Visa, it's unlikely you'll face such issues with proper member service support.

## Education Visa

If you're interested in learning something new while staying in Thailand, the Education Visa (ED Visa) might be the perfect fit. This visa is typically used by those enrolling in language, cooking, or other educational courses.

**Validity**: The Education Visa is generally valid for **90 days** initially, with the possibility of extensions every 90 days up to a year or more, depending on the course duration.

- **Requirements**: You must enrol in an accredited educational institution, such as a language school, and provide proof of enrolment.

**Types of education allowed**

- **Thai language courses**: One of the most common uses of the ED Visa. Many schools offer comprehensive Thai language programmes.
- **English language courses**: For non-native English speakers, enrolling in English language courses is also an option.
- **Muay Thai training**: Many accredited Muay Thai gyms offer long-term training programmes that qualify for an ED Visa.
- **Cooking schools**: If you're passionate about Thai cuisine, several cooking schools offer long-term courses that can sponsor an ED Visa.
- **Other courses**: Various other educational institutions offer courses in yoga, meditation, traditional Thai arts, and more.

**Pros**

- **Long-term stay**: Allows for an extended stay in Thailand while you study.

- **Learning opportunity**: Great for those who want to learn Thai or other skills.
- **Relatively simple process**: Schools often assist with the visa application process.

**Cons**

- **Study requirement**: You must attend classes regularly to maintain your visa status.
- **Limited work opportunities**: Officially, you're not allowed to work on an Education Visa.
- **Authority checks**: Schools can be visited by authorities to check attendance and ability. Immigration staff may quiz individuals on ED Visas, especially those studying Thai, to ensure they are genuinely studying and not using the visa as a cover to stay in Thailand illegally.

**Step-by-step guide to applying**

- **Enrol in a course**: First, you must enrol in an accredited educational institution in Thailand that offers the course you want to study.
- **Gather your documents**: Collect all necessary documents, including proof of enrolment, passport-sized photos, and financial evidence.
- **Get a letter of acceptance**: Obtain a letter of acceptance from the educational institution, which confirms your enrolment.
- **Fill out the application form**: Complete the visa application form, which is available on the embassy or consulate's website.
- **Submit your application**: Submit your application in person at the Thai embassy or consulate. Some consulates may allow for postal applications, but this varies.
- **Pay the visa fee**: Fees can vary depending on the embassy or consulate.

- **Wait for processing**: Processing times can vary, often taking a few days to a couple of weeks.
- **Collect your visa**: Once approved, collect your visa from the embassy or consulate.

**Necessary documents**

- **Passport**: A valid passport with at least six months remaining validity and at least one blank page.
- **Visa application form**: Available on the embassy or consulate's website.
- **Passport-sized photos**: Typically two photos, meeting the specific size requirements outlined by the embassy.
- **Proof of enrolment**: A letter of acceptance or proof of enrolment from an accredited educational institution in Thailand.
- **Proof of financial means**: Bank statements showing sufficient funds to cover your stay.
- **Proof of accommodation**: Hotel bookings or an invitation letter from a friend or family member in Thailand.

## How to extend your stay

### Prepare your documents

- **Passport**: Ensure your passport is valid for at least six months beyond your intended stay.
- **Visa extension application form (TM.7)**: Available at immigration offices or for download online.
- **Passport-sized photos**: Typically, two photos that meet the specifications outlined by the immigration office.
- **Proof of enrolment**: A letter from your educational institution confirming your continued enrolment.
- **Proof of financial means**: Bank statements showing sufficient funds to cover your stay.

- **Application fee**: The extension fee is **THB 1,900**, payable in cash.

**Visit the Immigration Office**

- Locate the nearest immigration office. In major cities like Bangkok and Chiang Mai, there are specific offices for this purpose.
- Arrive early to avoid long queues.

**Submit your application**

- Complete the **TM.7** form and attach your photos.
- Submit your application along with the required documents and fee.
- An immigration officer will review your application, and you may be asked a few questions about your studies.

**Wait for processing**

- Processing times can vary, but it usually takes a couple of hours. Some offices may ask you to return later in the day or the next day to collect your passport.

**Collect your passport**

- Once your extension is approved, collect your passport with the new visa stamp. You've successfully extended your stay!

**Important deadlines and procedures**

- **Extension deadline**: Apply for your extension at least two weeks before your current visa expires to allow sufficient processing time.
- **Overstay penalties**: Like other visas, overstaying an Education Visa incurs a fine of **THB 500 per day**, up to a maximum of **THB 20,000**. Overstays of more than 90

days can result in being blacklisted from re-entering Thailand for up to one year or more.

- **Attendance requirement**: Ensure you maintain regular attendance at your educational institution. Failure to do so can affect your visa extensions and renewals.

## Guardian Visa

The Guardian Visa is designed for parents or legal guardians (e.g. grandparents) of children enrolled in educational institutions in Thailand. This visa allows parents to stay in Thailand legally while their child attends school. This is an excellent and popular option for families.

**Validity**: The Guardian Visa is typically valid for 90 days initially, with the possibility of extensions every 90 days up to a year or more, depending on the child's enrolment status.

- **Requirements**: You must provide proof of your child's enrolment in an accredited educational institution in Thailand. This paperwork can be obtained quite easily from your educational institution of choice. The only requirements are that the child is between 3 and 18 years old and enrolled full-time. However, only one Guardian Visa is allowed per child, so two parents would require two children to make the most of this visa opportunity.

**Pros**

- **Family unity**: Allows parents to stay with their children while they study in Thailand.
- **Extension possibility**: Can be extended multiple times as long as the child remains enrolled in school.
- **Supportive environment**: Ensures that children have parental support while adapting to a new educational environment.

**Cons**

- **Limited work opportunities**: Officially, you're not allowed to work on a Guardian Visa.
- **Documentation requirements**: Requires substantial documentation, including proof of the child's enrolment, guardianship, and financial means to support your stay.
- **Dependent on child's status**: The visa is tied to the child's enrolment status, so any changes in the child's schooling can affect the visa.

**Step-by-step guide to applying**

- **Enrol your child**: Ensure your child is enrolled in an accredited educational institution in Thailand.
- **Gather your documents**: Collect all necessary documents, including proof of your child's enrolment, passport-sized photos, and financial evidence.
- **Get a letter from the school**: Obtain a letter from the educational institution confirming your child's enrolment and your status as a guardian.
- **Fill out the application form**: Complete the visa application form, available on the embassy or consulate's website.
- **Submit your application**: Submit your application in person at the Thai embassy or consulate. Some consulates may allow for postal applications, but this varies.
- **Pay the visa fee**: Fees can vary depending on the embassy or consulate.
- **Wait for processing**: Processing times can vary, often taking a few days to a couple of weeks.
- **Collect your visa**: Once approved, collect your visa from the embassy or consulate.

**Necessary documents**

- **Passport**: A valid passport with at least six months remaining validity and at least one blank page.
- **Visa application form**: Available on the embassy or consulate's website.
- **Passport-sized photos**: Typically two photos, meeting the specific size requirements outlined by the embassy.
- **Proof of child's enrolment**: A letter from the accredited educational institution in Thailand confirming your child's enrolment.
- **Proof of financial means**: Bank statements showing sufficient funds to cover your stay.
- **Proof of accommodation**: Hotel bookings or a lease agreement.
- **Proof of guardianship**: Documents proving your relationship to the child, such as a birth certificate or legal guardianship papers.

## How to extend your stay

**Prepare your documents**

- **Passport**: Ensure your passport is valid for at least six months beyond your intended stay.
- **Visa extension application form (TM.7)**: Available at immigration offices or for download online.
- **Passport-sized photos**: Typically, two photos that meet the specifications outlined by the immigration office.
- **Proof of child's enrolment**: A letter from the educational institution confirming your child's continued enrolment.
- **Proof of financial means**: Bank statements showing sufficient funds to cover your stay.
- **Application fee**: The extension fee is **THB 1,900**, payable in cash.

**Visit the Immigration Office**

- Locate the nearest immigration office. In major cities, there are specific offices for this purpose.
- Arrive early to avoid long queues.

**Submit your application**

- Complete the **TM.7** form and attach your photos.
- Submit your application along with the required documents and fee.
- An immigration officer will review your application, and you may be asked a few questions about your child's schooling.

**Wait for processing**

- Processing times can vary, but it usually takes a couple of hours. Some offices may ask you to return later in the day or the next day to collect your passport.

**Collect your passport**

- Once your extension is approved, collect your passport with the new visa stamp. You've successfully extended your stay!

**Important deadlines and procedures**

- **Extension deadline**: Apply for your extension at least two weeks before your current visa expires to allow sufficient processing time.
- **Overstay penalties**: Overstaying a Guardian Visa incurs the same penalties as other visas, with fines of **THB 500 per day**, up to a maximum of **THB 20,000**. Longer overstays can lead to being blacklisted.
- **Child's enrolment**: Your visa extension depends on your child's continued enrolment in an accredited educational institution. Ensure that all necessary documentation is up-to-date and submitted on time.

## 90-DAY REPORTING

Regardless of the type of visa you hold, if you stay in Thailand for an extended period, you are required to report your address to the Thai Immigration Bureau every 90 days. This process, known as '90-day reporting', is a crucial part of maintaining your legal status in the country. Failure to comply with this requirement can result in fines or other legal issues.

### Staying On the Right Side of the Law

#### What is 90-day reporting?

90-day reporting is a requirement for all foreign nationals staying in Thailand for more than 90 consecutive days. It ensures that the Thai government has up-to-date information on the whereabouts of long-term residents. This requirement is separate from your visa validity or extensions. The process must be completed every 90 days if you remain in the country continuously. The reporting can be done in person, online, or by post, making it relatively convenient. It is crucial to keep track of your reporting dates to avoid any issues with Thai immigration authorities.

**Note:** Unlike other visas, **Smart Visa** holders are required to report their stay to immigration only once a year rather than every 90 days. However, it's essential to stay on top of this annual reporting to avoid any legal issues. Ensure you mark your calendar and set reminders for this important date.

#### Why is it required?

The 90-day reporting requirement helps Thai authorities keep track of foreign nationals within the country. It is part of their immigration control measures to ensure security and proper management of foreign residents. By maintaining up-to-date records of where foreign nationals are residing, the government can more effectively manage and monitor the movement of people within the country. This is particularly important for national security, public safety, and resource planning.

Additionally, regular reporting helps authorities identify and address any potential legal or immigration issues promptly. It ensures that foreign nationals comply with Thai laws and visa conditions, thereby reducing the risk of illegal activities and overstays. For residents, adhering to the 90-day reporting rule demonstrates a commitment to abiding by local regulations, which can benefit future visa extensions or applications.

**Consequences of non-compliance**

Failing to comply with the 90-day reporting requirement can result in penalties:

- **Fines**: If you miss your reporting date, you can be fined **THB 2,000**. If authorities catch you before you report, the fine may increase to **THB 4,000**. These fines can quickly add up if you repeatedly miss reporting deadlines.
- **Legal issues**: Repeated non-compliance can lead to more severe consequences, including potential detention and impact on future visa extensions or applications. Authorities may view repeated failure to report as a sign of disregard for Thai laws, which can result in stricter scrutiny and difficulties securing long-term residency.
- **Deportation**: In extreme cases, non-compliance with the 90-day reporting requirement, especially if coupled with other violations, can result in deportation. This can significantly disrupt your stay in Thailand and adversely affect your ability to return.
- **Blacklisting**: Chronic offenders may be blacklisted, preventing them from re-entering Thailand for a specified period. This can severely restrict your travel plans and future opportunities in the country.

- **Reputation damage**: Non-compliance can damage your reputation with Thai authorities, making it more challenging to navigate other bureaucratic processes smoothly. Maintaining a clean record demonstrates respect for local regulations and can facilitate smoother interactions with immigration officials.

Staying on top of the 90-day reporting requirement is crucial for maintaining your legal status in Thailand and ensuring a trouble-free stay. Always set reminders and plan ahead to meet your reporting obligations.

**How to do 90-day reporting**

There are several ways to complete your 90-day reporting, making it a relatively convenient process:

**In-person**

- **Visit the nearest immigration office**: You can report in person within **15 days before your reporting date and up to 7 days after**. It's advisable to complete the process as early as possible within this window to avoid any issues.
- **Bring your documents**: Bring your passport, completed **TM.47 form** (available at the office or online), and a copy of your passport ID page, current visa page, and the last entry stamp page.
- **Submit your documents**: Submit your documents to the immigration officer. You will receive a receipt with the next reporting date.

**Online**

- **Note**: Online reporting is only available if you have previously completed an in-person 90-day report.
- **Visit the Thai Immigration Bureau's official website**: Navigate to the 90-day reporting section. The same

reporting window as in person applies – **15 days before your reporting date and up to 7 days after**.

- **Complete the online form**: Complete the online form and upload scanned copies of your passport pages and the TM.47 form.
- **Submit your report online**: If successful, you will receive a confirmation email with the next reporting date.

**By post**

- **Prepare your documents**: Send your TM.47 form, copies of the relevant passport pages, and a self-addressed stamped envelope to your local immigration office at least **15 days before your reporting date**.
- **Use registered mail**: Ensure you use registered mail to track your documents.
- **Receive a receipt by post**: You will receive a receipt by post with the next reporting date.

**Conclusion**

Staying on top of your 90-day reporting is essential for maintaining your legal status in Thailand. Whether you opt to report in person, online, or by post, make sure to do so on time to avoid penalties. Non-compliance can result in fines or more severe consequences, such as detention or issues with future visa applications. By keeping track of your reporting dates and setting reminders, you can ensure a seamless and worry-free stay in Thailand. Proper adherence to Thai immigration regulations not only maintains your legal status but also contributes to a positive experience in the country.

## LEGAL CONSIDERATIONS

So, you've got your visa sorted and have even extended your stay. But as a digital nomad, there's another piece of the puzzle you need to understand: work permits and legal considerations.

Thailand's regulations around work can be a bit tricky, but with the right knowledge, you can navigate them smoothly and avoid any legal issues. In this section, we'll explore the ins and outs of obtaining a work permit, staying legal, and avoiding common pitfalls.

## Understanding Work Permits for Digital Nomads

One of the most important legal considerations for digital nomads in Thailand is understanding whether you need a work permit and how to obtain one. Here's a comprehensive look at what you need to know.

### Do you need a work permit?

The question of whether digital nomads need a work permit in Thailand is a bit of a grey area. Officially, anyone engaging in work, whether paid or unpaid, physical or mental, requires a work permit. This includes online work, regardless of whether the work is for a company based outside of Thailand. The Thai government broadly defines 'work' as any activity that provides a benefit, whether financial or otherwise.

However, enforcement can be inconsistent, and many digital nomads work remotely without a permit. It's crucial to weigh the risks and make an informed decision. Consider your long-term plans and the nature of your work. If your activities involve local clients or frequent business interactions within Thailand, obtaining a work permit is advisable.

On the other hand, if your work is entirely online and you have no physical presence in the Thai business community, you might be less likely to encounter enforcement. Consulting with a legal expert or an immigration professional can provide valuable insights and help you navigate this complex issue.

## Types of work permits available

Several types of work permits are available, depending on your situation. Understanding the options can help you choose the most suitable permit for your needs.

- **Traditional work permit**: Typically tied to a specific job with a Thai company. This type of permit requires sponsorship from your employer, who must provide various company documents and an employment contract. It is suited for those who have secured a job offer from a Thai business.
- **Smart visa**: Designed for entrepreneurs, investors, skilled professionals, and startups in specific industries such as tech and innovation. The Smart Visa allows you to work without a traditional work permit and offers additional benefits like longer stays and fewer reporting requirements. This visa is ideal for those involved in innovative and high-growth sectors.
- **Business visa with work permit**: If you set up your own company in Thailand, you can apply for a business visa and a corresponding work permit. This option requires you to establish a legal entity in Thailand, which involves meeting specific financial and operational criteria. It is perfect for entrepreneurs looking to base their operations in Thailand.

Each type of work permit has its own advantages and requirements, so it's essential to assess your specific situation and long-term goals before applying.

## Applying for a work permit

If you decide to go the legal route and obtain a work permit, here's a step-by-step guide:

**Obtain the correct visa:** Before applying for a work permit, you must have the appropriate visa, usually a Non-Immigrant B Visa.

- **Find a sponsor:** If you're employed by a Thai company, they will act as your sponsor. If you're self-employed, you must set up a Thai company* to sponsor yourself.
- **Prepare your documents:**
- **Passport:** A valid passport with at least six months of validity remaining.
- **Non-Immigrant visa:** A valid Non-Immigrant B Visa.
- **Work permit application form:** Available at the Ministry of Labour or for download online.
- **Passport-sized photos:** Typically three photos, meeting the specifications outlined by the Ministry of Labour.
- **Medical certificate:** A health certificate from a Thai hospital or clinic.
- **Company documents:** For employees, this includes the company's registration documents, list of shareholders, VAT registration, and financial statements. Self-employed individuals will need their company's registration documents.
- **Employment contract:** A signed contract outlining your job role and responsibilities.
- **Submit your application:** Submit your application and documents to the Ministry of Labour. There is an application fee, which varies depending on the length of the work permit.
- **Attend an Interview:** You may be required to attend an interview at the Ministry of Labour.
- **Wait for processing:** Processing times can vary, but it usually takes a few weeks.
- **Collect your work permit:** Once approved, you can collect your work permit from the Ministry of Labour.

**Note:** Setting up a Thai company is a rigorous process, which is beyond the scope of this book. Unless you plan to stay in Thailand long-term or particularly want to set up a local business, we don't recommend this option due to its complexity, expense, and upkeep (e.g. annual accounting and legal requirements). Furthermore, even if you set up a company that does not automatically grant you the right to stay in Thailand, you still need to apply for a visa and work permit.

## Special Considerations for Education Visa and Guardian Visa Holders

While Education Visa and Guardian Visa holders are generally not permitted to work in Thailand, specific legal considerations must be considered to ensure compliance with Thai immigration laws.

### Education Visa holders

**Study requirement:** Maintain regular attendance at your educational institution. Immigration authorities can check your attendance records and quiz you on your studies to ensure you genuinely study.

- **Non-work status:** Officially, you cannot work on an Education Visa. Working without a permit can result in fines, detention, and deportation.
- **Compliance:** Ensure all your documentation is up-to-date, including proof of enrolment and financial means.

### Guardian Visa holders

- **Child's enrolment:** Your visa is tied to your child's continued enrolment in an accredited educational institution. Ensure that your child's enrolment status is maintained and documented.

- **Non-work status:** Officially, you cannot work on a Guardian Visa. Working without a permit can result in fines, detention, and deportation.
- **Compliance:** Be prepared to provide comprehensive documentation proving your relationship to the child and financial means to support your stay.

## Staying Legal and Avoiding Common Pitfalls

Navigating the legal landscape as a digital nomad in Thailand can be tricky, but here are some tips to help you stay on the right side of the law and avoid common pitfalls.

### Staying out of trouble

**Be informed and stay updated**: Visa and work permit regulations can change, so it's crucial to stay informed about the latest requirements and updates. Regularly check the official websites of the Thai Immigration Bureau and the Ministry of Labour for any changes.

**Avoid overstaying**: Overstaying your visa can lead to hefty fines, detention, and even being blacklisted from re-entering Thailand. Always keep track of your visa's expiry date and apply for extensions or renewals well in advance.

**Be cautious with visa runs**: While visa runs (leaving Thailand and re-entering to reset your visa) are a common practice, they come with risks. Immigration officers may deny entry if they suspect you are abusing the system. It's better to apply for a proper visa extension or change your visa type if you plan to stay long-term.

**Understand the implications of working without a permit**: Working without a permit is illegal in Thailand and can result in fines, detention, and deportation. If you're caught working without a permit, you could face serious consequences. Weigh the risks carefully and consider obtaining a work permit if you plan to work long-term in Thailand.

**Keep your documents organised**: Always keep copies of your visa, work permit, passport, and other important documents. Organised and easily accessible documents can save you a lot of hassle during inspections or if you need to renew or extend your visa.

**Seek professional advice**: Navigating Thai immigration and labour laws can be complex. If you're unsure about any aspect of your visa or work permit, consider seeking advice from a professional immigration lawyer or a reputable visa agency. They can provide valuable guidance and ensure you comply with all legal requirements.

**Respect Thai laws and customs**: Thailand is known for its rich culture and traditions. Respect local laws, customs, and etiquette. Avoid engaging in activities that could be considered disrespectful or illegal, such as drug use or disrespecting the monarchy.

**For Education Visa holders**: Maintain regular attendance at your educational institution. Immigration authorities can check your attendance records and quiz you on your studies to verify that you are genuinely studying. Non-compliance can lead to visa revocation.

**For Guardian Visa holders**: Ensure your child's enrolment status is maintained and documented. Your visa is tied to your child's continued enrolment in an accredited educational institution. Be prepared to provide updated enrolment confirmation and financial documentation during each extension.

**Conclusion**

Navigating the legal considerations of living and working in Thailand as a digital nomad might seem daunting, but with the right information and preparation, it's entirely manageable. Understanding work permits, staying informed about visa regulations, and adhering to local laws are key to enjoying a hassle-free stay in the Land of Smiles. By keeping your documents organised, seeking professional advice when needed, and respecting Thai customs and laws, you can avoid common pitfalls and ensure a smooth experience. With these precautions, you can fully enjoy your time in Thailand, focusing on your work and the incredible experiences the country offers.

**Stay tuned . . .**

Stick around because in **Chapter 5: Health & Wellness**, we'll dive into maintaining your health and wellness while living in Thailand. We'll explore the healthcare system, essential health insurance options for expats, and ways to stay fit and active through fitness centres, yoga studios, and outdoor activities. Understanding the local healthcare landscape will help you make informed decisions about your health, ensuring access to quality medical care whenever needed.

We'll also address common health concerns, such as necessary vaccinations and dealing with food and water safety, ensuring you stay healthy and thrive during your time in Thailand. Additionally, we'll cover mental health resources, traditional Thai medicine, and wellness retreats that can enrich your experience. By prioritising your health and wellness, you can fully enjoy the vibrant lifestyle that Thailand offers. So, stay tuned for practical tips and insights to help you live a balanced and healthy life in the Land of Smiles.

# Chapter 5:
# Health & Wellness

### Staying Fit and Healthy in the Land of Smiles

Welcome to the chapter that keeps you in tip-top shape while living your dream in Thailand! Navigating a new healthcare system can be daunting but fear not – we've got you covered. From finding the best hospitals and clinics to comparing health insurance options, we'll ensure you're well-prepared. Plus, we'll dive into the vibrant fitness scene, from gyms to yoga studios, and offer tips on staying healthy in a tropical paradise. We'll also explore traditional Thai wellness practices and local health trends. Ready to discover how to maintain your wellness in the Land of Smiles? Let's get started!

## COMMON HEALTH CONCERNS

Living in a tropical paradise like Thailand comes with its own set of health considerations. From ensuring you have the right vaccinations to staying healthy amidst the heat and humidity, it's important to be prepared. This section will guide you through the necessary vaccinations and provide tips for maintaining your health in a tropical climate. We'll cover common health issues such as mosquito-borne diseases, food and water safety, and how to manage the effects of heat and humidity. Following these guidelines, you can enjoy your time in Thailand while staying fit and healthy.

### Essential Vaccinations

Before travelling to Thailand, ensure your routine vaccinations are up to date so you can stay in tip-top shape throughout your stay. Of course, the unexpected can happen, so we list below some other common vaccinations you should consider getting before arriving in the Land of Smiles.

## What jabs to get and when

### Routine immunisations

- Measles, mumps and rubella (MMR)
- Diphtheria, tetanus and pertussis (DTaP)
- Varicella (chickenpox)
- Polio
- Annual flu shot

These vaccinations are typically recommended for everyone, regardless of travel plans, to protect against common diseases.

### Hepatitis A and B

Hepatitis A and B are viral infections that affect the liver. Hepatitis A is usually contracted through contaminated food and water, while Hepatitis B is spread through contact with infected blood or bodily fluids. Vaccinations for both Hepatitis A and B are strongly recommended for travellers to Thailand.

- **Hepatitis A:** The vaccine is given in two doses, six months apart.
- **Hepatitis B:** The vaccine is given in three doses over six months.

### Typhoid

Typhoid fever is caused by the bacterium Salmonella typhi and is contracted through contaminated food and water. A typhoid vaccine is recommended for travellers to Thailand, especially if you plan to visit rural areas or eat street food. The vaccine is available as an injection or oral capsule.

- **Injection:** One dose provides protection for two years.
- **Oral capsules:** Four capsules taken over a week provide protection for five years.

### Japanese encephalitis

Japanese encephalitis is a mosquito-borne viral infection that can cause severe brain inflammation. The risk is higher in rural areas and during the rainy season. The vaccine is recommended for travellers who plan to spend extended periods in rural areas or participate in outdoor activities.

- **Vaccine:** Two doses are given 28 days apart, with a booster dose recommended after one year.

### Rabies

Rabies is a viral infection transmitted through the bite or scratch of an infected animal, such as dogs, bats or monkeys. Although the risk is low, the consequences of rabies are severe. The vaccine is recommended for travellers who may have contact with animals or are visiting remote areas where medical care may be limited.

- **Pre-exposure vaccine**: Three doses are given over 28 days.

### Malaria prevention

Malaria is a mosquito-borne disease prevalent in some parts of Thailand, particularly in rural and forest areas. While there is no vaccine for malaria, preventive measures and medications can help reduce the risk.

- **Antimalarial medication:** Consult your doctor for the appropriate medication based on your travel itinerary. Common options include doxycycline, atovaquone-proguanil and mefloquine.
- **Preventive measures:** Use insect repellent, wear long sleeves and pants and sleep under mosquito nets to minimise exposure to mosquito bites.

### Dengue fever

Dengue fever is another mosquito-borne illness common in Thailand, especially during the rainy season. There is currently no widely available vaccine for dengue, so prevention is crucial.

- **Preventive measures:** Use mosquito repellent, wear long sleeves and pants and stay in accommodations with screened windows and doors. Consider using mosquito nets while sleeping.

### Covid-19

Of course, we would be remiss if we didn't mention Covid-19. The global pandemic has made it essential to stay updated on the latest information and guidelines regarding Covid-19 vaccinations and safety measures. Here's what you need to know:

- **Covid-19 vaccination:** Ensure you are fully vaccinated against Covid-19 before travelling to Thailand. Many countries now require proof of vaccination for entry. Booster doses may also be recommended depending on the guidelines during your travel.
- **Local guidelines:** Stay informed about Thailand's Covid-19 regulations, including mask mandates, social distancing measures and travel restrictions. Regulations can change frequently, so it's important to keep up to date with official sources such as the Thai Ministry of Public Health.
- **Health precautions:** Continue practising good hygiene, such as frequent hand washing, using hand sanitiser and wearing masks in crowded places. If you experience symptoms of Covid-19, seek medical attention immediately and follow local health guidelines for testing and isolation.

- **Travel insurance:** Consider purchasing travel insurance that covers Covid-19-related medical expenses and trip cancellations. This can provide peace of mind and financial protection in case of unexpected disruptions to your travel plans.

## Staying Healthy in a Tropical Climate

### Drinking water and diarrhoea

Staying hydrated is crucial in a tropical climate. The heat and humidity can cause you to sweat more, increasing your risk of dehydration. However, it's important to ensure your drinking water is safe. Here are some tips to help you avoid waterborne illnesses:

- Bottled water: The safest option for drinking water in Thailand is bottled water. Always check that the seal is intact before purchasing. Bottled water is widely available and inexpensive, costing around **THB 10-20 per bottle**. Many hotels and restaurants provide complimentary bottled water for guests.
- Filtered water: Some hotels, restaurants and homes use water filtration systems to provide safe drinking water. If you have access to filtered water, it's generally safe to drink. However, it's a good idea to confirm the quality of the filtration system before relying on it exclusively.
- Avoid tap water: Tap water in Thailand is unsafe for drinking. Avoid using tap water for drinking, brushing your teeth or making ice. Stick to bottled or filtered water to prevent waterborne illnesses.
- Ice safety: Ice can be a source of contamination if it's made from tap water. When ordering drinks, ask if the ice is made from filtered or bottled water. If you're unsure, it's safer to avoid ice altogether.

Traveller's diarrhoea is a common ailment among visitors to Thailand, often caused by consuming contaminated food or water. Symptoms include stomach cramps, diarrhoea and sometimes fever. To prevent traveller's diarrhoea, follow the food and water safety tips outlined above.

- **Prevention:** Maintain good hygiene, choose reputable food vendors and drink bottled or filtered water.
- **Treatment:** Stay hydrated, rest and consider over-the-counter medications such as loperamide. If symptoms persist or worsen, seek medical attention.
- **Probiotics:** Consider taking probiotics either before or during your trip to help maintain a healthy balance of gut bacteria, which can reduce the risk of traveller's diarrhoea.

**Food safety and poisoning**

Thailand is famous for its delicious street food, but it's essential to be cautious about food safety. Follow these tips to reduce the risk of foodborne illnesses:

**Choose busy food stalls:** Busy stalls with high turnover are likelier to have fresh food.

- **Watch the preparation:** Ensure food is cooked thoroughly and served hot. Avoid raw or undercooked dishes.
- **Peel fruits and vegetables:** If you're unsure about the cleanliness of produce, stick to fruits and vegetables that can be peeled. Always wash any produce you buy before eating.
- **Avoid ice:** Ice made from tap water can carry bacteria. Opt for drinks without ice or ask if the ice is made from filtered water.

Not surprisingly, food poisoning can result from consuming contaminated food or drinks. Symptoms include nausea, vomiting, diarrhoea and abdominal pain. To avoid food poisoning, practice safe food handling and preparation, and be cautious when dining out.

- **Prevention:** Follow proper food safety practices, avoid raw or undercooked foods and ensure food is cooked thoroughly.
- **Treatment:** Stay hydrated, get plenty of rest and seek medical advice if symptoms are severe or prolonged.
- **When to seek medical attention:** If you experience high fever, blood in your stool, or prolonged symptoms lasting more than 48 hours, it's crucial to seek medical attention immediately. These could be signs of a more serious infection that requires professional treatment.

**Sun protection and heat exposure**

The tropical sun in Thailand can be intense, so it's important to protect your skin from harmful UV rays. Use a broad-spectrum sunscreen with an SPF of at least 30 and reapply it every two hours, especially if you're swimming or sweating. Wear a wide-brimmed hat and sunglasses to shield your face and eyes from the sun.

- **Rash vests:** Consider wearing rash vests or UV-protective clothing, especially during extended periods of outdoor activities. These garments provide additional protection against UV rays and can be more effective than sunscreen alone.
- **Stay under shade:** Seek shade whenever possible, especially during peak sunlight hours (usually between 11am and 3pm). Using an umbrella, canopy or finding natural shade can significantly reduce your exposure to UV rays.
- **Avoid peak sun hours:** Limit your time in direct sunlight during the hottest part of the day. Plan outdoor

activities for early morning or late afternoon when the sun's rays are less intense.

- **Hydration:** Staying hydrated is also crucial as it helps your skin stay healthy and better protected against sun damage. Drink plenty of water throughout the day.

Heat exhaustion and heatstroke are serious conditions in hot and humid climates. To prevent these, follow these tips:

**Stay cool:** Avoid strenuous activities during the hottest part of the day (usually between 11pm and 3pm).

- **Stay hydrated:** Drink plenty of water and avoid excessive caffeine or alcohol, which can dehydrate you.
- **Listen to your body:** If you start to feel dizzy, weak, or nauseous, find a cool place to rest and drink water.

**Insect prevention**

Mosquitoes in Thailand can carry diseases such as dengue fever, malaria, and Zika virus. Protect yourself from insect bites with these measures:

- **Use insect repellent:** Apply a repellent containing DEET, picaridin or oil of lemon eucalyptus to exposed skin.
- **Wear protective clothing:** Opt for long sleeves, long pants and closed-toe shoes, especially during dawn and dusk when mosquitoes are most active.
- **Sleep under mosquito nets:** If you're staying in an area with a high mosquito population, use a mosquito net over your bed.
- **Other insects:** Be cautious of other insects such as sand flies, spiders, and scorpions. Use insect repellent and wear protective clothing to avoid bites and stings. If bitten or stung, seek medical attention if there are any signs of a severe reaction.

**Conclusion**

Staying fit and healthy as a digital nomad in Thailand is about being prepared and making informed choices. By understanding the healthcare system, securing the right health insurance and navigating food and water safety, you can enjoy the vibrant culture and beautiful landscapes with peace of mind. Staying active through fitness centres, yoga studios, and outdoor activities will keep your body and mind in top shape. Remember, your health is your greatest asset, so take the necessary steps to protect it while living your dream in Thailand. Keep thriving and stay healthy in the Land of Smiles!

# HEALTHCARE SYSTEM IN THAILAND

When it comes to healthcare, Thailand offers a range of options that cater to both expats and locals. Whether you prefer the affordability of the public system or the higher-end services of private hospitals, knowing your choices can make all the difference. Let's dive into the details to help you navigate the local healthcare landscape easily and confidently.

## Overview of Available Healthcare Options

### Public healthcare system

### Affordability and accessibility

The public healthcare system in Thailand is renowned for its affordability. For those holding a Thai work permit or social security card, public hospitals provide medical services at significantly reduced rates. Even without these benefits, the costs are generally much lower than in many Western countries. For example, a basic consultation at a public hospital may cost around THB 50-100, and a more specialised visit might be around THB 200-500. Medicines are also cheaper, usually costing less than THB 1,000.

### Quality of care

While the public healthcare system is affordable, the quality of care can vary. Major cities like Bangkok, Phuket, and Chiang Mai boast public hospitals with modern facilities and highly trained staff. However, in more rural areas, the quality may not be as high, and language barriers can be an issue, as not all medical staff speak English fluently.

### Wait times

One downside of public healthcare is the wait times. Given the high patient volume, you might find yourself waiting several hours for a consultation, especially if they're scrambling around to find an English-speaking medical team. It's advisable to arrive early and bring something to keep you occupied. Patience is key!

### Services offered

Public hospitals offer various medical services, from emergency care to routine check-ups and specialist consultations. Some hospitals also have specialised departments for cardiology, orthopaedics, cancer, and other fields. However, for more complex or urgent medical needs, many expats prefer to use private facilities.

## Private healthcare system

### Quality and convenience

Private hospitals in Thailand are often compared to high-end hotels, offering a level of comfort and service that can make medical visits almost pleasant. These hospitals are known for their state-of-the-art facilities, well-trained staff, and a focus on patient comfort. Many private hospitals are accredited by international organisations, ensuring a high standard of care.

### Language and communication

One of the significant advantages of private hospitals is the ease of communication. Most doctors and medical staff in private hospitals speak English fluently, and some facilities even offer services in other languages like Chinese, Japanese, and Arabic. This can be a huge relief if you're not confident in your Thai language skills. Other translation services are also available upon request, usually requiring advance notice.

### Costs

As you might expect, the superior services and facilities come with a higher price tag. A basic consultation at a private hospital can range from THB 500 to 1,500, while more specialised treatments can cost significantly more. Despite the higher costs, many expats find the quality of care worth the investment, especially for serious or complex medical issues. In the mid-range section, many private clinics offer English services for everyday illnesses, although they may not provide specialised care.

### Insurance coverage

Many private hospitals accept international health insurance plans, making it easier for expats to get the care they need without worrying about upfront costs. It's advisable to check with your insurance provider to see which hospitals are covered under your plan. Some popular private hospital chains include Bumrungrad International Hospital, Bangkok Hospital, and Samitivej Hospital. Show your insurance card when you 'check in' to ensure your visit is covered (e.g. no upfront payment required). However, coverage can vary on a case-by-case basis, and sometimes a deposit is required to receive treatment while waiting for the health insurance company to confirm payment.

### Specialist care

Private hospitals often have a wider range of specialists available, from dermatologists to cardiologists, ensuring you can get the expert care you need. These specialists are frequently trained internationally and bring extensive experience to their fields. Private hospitals also tend to have more advanced diagnostic and treatment facilities, such as MRI machines and specialised surgical equipment, which can make a significant difference in the quality of care. Appointments are usually quicker and more flexible than in public hospitals, allowing you to schedule visits that fit your lifestyle. Additionally, many private hospitals offer telemedicine consultations, providing you with the convenience of expert advice without leaving your home. Whether you need routine check-ups or specialised treatments, the comprehensive range of services offered by private hospitals ensures you're well taken care of.

### Cosmetic surgery

Many international hospitals now offer an 'aesthetics' department, performing various procedures, from rhinoplasty (nose jobs) and breast augmentation to facelifts and liposuction. Thailand has become a popular destination for medical tourism, particularly for cosmetic surgery, due to the high quality of care and competitive pricing. Many of these procedures are performed by highly skilled surgeons with international training and experience. The hospitals provide comprehensive packages that often include consultations, surgery, aftercare, and even accommodation for recovery. It's advisable to research and choose accredited facilities and qualified surgeons to ensure the best outcomes.

## Finding local clinics and how to pick the right one

When it comes to healthcare in Thailand, both private and public clinics offer a range of services. Here's how to find and choose the right one for you:

### Proximity and convenience

Look for clinics near your home or workplace to make visits easier and less time-consuming. Google Maps can help you locate nearby clinics and check their reviews.

### Specialist clinics

Thailand has numerous specialist clinics for dental care, dermatology, and physiotherapy. Research clinics that focus on your specific health needs to ensure you receive expert care. Websites like WhatClinic can help you find and compare specialist clinics.

### General practices

For general healthcare needs, there are many reputable general practice clinics throughout Thailand. These clinics offer services like routine check-ups, vaccinations, and minor treatments. Some well-known general practice chains include BNH Clinic and MedConsult.

### Visit in person

Whenever possible, visit the clinic in person before deciding. This allows you to assess cleanliness, facilities, and the overall environment. Pay attention to how the staff interact with patients and whether the clinic feels welcoming and professional.

### Check language capabilities

Ensure the clinic has staff who speak your language fluently, or at least English, to avoid misunderstandings and ensure clear communication about your health needs.

**Conclusion**

Finding a good hospital or clinic in Thailand doesn't have to be a daunting task. By leveraging online resources, seeking recommendations from the expat community, and visiting facilities in person, you can ensure you receive high-quality care. Whether you choose a renowned private hospital or a convenient local clinic, Thailand offers a range of excellent healthcare options. Additionally, understanding the local healthcare landscape and knowing your insurance options can further ease the process. Stay proactive in your healthcare choices and enjoy a fit and healthy life in the Land of Smiles. Here's to your well-being!

## HEALTH INSURANCE OPTIONS

Living in a foreign country brings many uncertainties, and health-related issues can add to the stress. Having reliable health insurance provides peace of mind, knowing that you're covered in case of emergencies, routine check-ups, or unexpected illnesses. Embarking on the digital nomad journey in Thailand is exhilarating, but ensuring you have the right health insurance is crucial to keep those adventures worry-free. With a plethora of options available, finding the right insurance can feel overwhelming. Fear not! This guide will help you navigate the maze of health insurance options, highlighting must-have coverage and tips for making the best choice for your needs.

### Types of Health Insurance Available

Choosing the right health insurance ensures access to quality medical care in Thailand. As an expat or digital nomad, you have two main options: local and international health insurance. Each type has advantages and disadvantages, making it important to carefully consider your needs and travel habits.

Below, we'll explore the specifics of local and international health insurance plans to help you make an informed decision.

**Local health insurance**

Local health insurance plans are provided by Thai insurance companies and are tailored to the needs of residents and expats in Thailand. These plans often cover various medical services, including hospitalisation, outpatient care, and sometimes dental and vision care.

**Pros**

- Generally more affordable than international plans.
- Designed specifically for the Thai healthcare system.
- Easy to purchase and manage with local providers.
- Doesn't offer a lot of perks, added extras, or incentives.

**Cons**

- Coverage may be limited to Thailand, which could be a drawback if you travel frequently.
- Some plans may have lower coverage limits and fewer benefits compared to international plans.
- Language barriers can sometimes be an issue when dealing with customer service or medical staff who may not speak fluent English.
- Not supported by an app or website in English.

**International health insurance**

International health insurance plans are designed for expats and offer worldwide coverage. These plans are ideal for digital nomads who travel frequently and need comprehensive coverage across multiple countries.

**Pros**

- Offers extensive coverage worldwide, including medical evacuation and repatriation.

- High coverage limits and a wide range of benefits.
- Flexibility to receive care in various countries.
- Comes with a convenient website and app in English for making claims and checking inclusions.

**Cons**

- Typically more expensive than local plans.
- May require more paperwork and documentation.
- Some plans might have exclusions or limitations for pre-existing conditions, affecting coverage.
- Usually excludes the United States.

## Must-have coverage components

### Inpatient and outpatient care

Ensure your plan covers inpatient (hospitalisation) and outpatient (doctor visits, specialist consultations) care. Inpatient care covers costs related to hospital stays, surgeries, and treatments, while outpatient care covers routine check-ups, diagnostics, and minor treatments.

### Emergency medical evacuation

In case of severe medical emergencies that require specialised care unavailable in Thailand, emergency medical evacuation coverage ensures you can be transported to a facility where you can receive the necessary treatment.

### Repatriation of mortal remains

While a sombre topic, it's important to have coverage for repatriating mortal remains. This ensures that, in the unfortunate event of death, your remains can be transported back to your home country.

### Dental and vision care

These are often optional add-ons but can be invaluable. Routine dental check-ups and vision care are essential for overall health, and having coverage for these can save you from out-of-pocket expenses.

**Preventive care**

Look for plans that cover preventive care, such as vaccinations, annual health check-ups, and screenings. Preventive care can help catch potential health issues early, leading to better outcomes and lower treatment costs.

## Top Insurance Providers for Expats in Thailand

### Local providers

**Allianz Ayudhya**

Allianz Ayudhya offers a range of health insurance plans tailored for expats. Their plans cover inpatient and outpatient care, emergency medical evacuation, and optional dental and vision care. They have a strong network of hospitals and clinics across Thailand.

**Luma Health Insurance**

Luma offers comprehensive health insurance plans with a focus on providing high-quality care for expats. Their plans include extensive inpatient and outpatient coverage, emergency evacuation, and preventive care. Luma is known for its excellent customer service and easy claims process.

**AXA Thailand**

AXA offers a variety of health insurance plans suitable for expats, including options for inpatient, outpatient, and specialist care. They provide comprehensive coverage and have a broad network of partner hospitals and clinics.

**Thai Health Insurance**

Thai Health Insurance provides a range of plans tailored for expats, covering inpatient and outpatient services, dental care, and emergency medical evacuation. They offer competitive pricing and have a network of accredited hospitals and clinics throughout Thailand.

**Pacific Cross Health Insurance**

Pacific Cross Health Insurance is well-regarded for its comprehensive health insurance plans catering to expats. They provide extensive coverage for inpatient and outpatient care, emergency evacuation, and preventive services. Known for their reliable customer service, they have a strong presence in the local healthcare market.

**International providers**

**Cigna Global**

Cigna Global is a well-known provider of international health insurance, offering flexible plans that can be customised to meet your needs. Their plans include inpatient and outpatient care, emergency evacuation, dental and vision care, and wellness programmes.

**Bupa Global**

Bupa Global offers extensive health insurance plans with worldwide coverage. Their plans include a wide range of benefits, such as inpatient and outpatient care, preventive care, emergency evacuation, and mental health support. Bupa is known for its high coverage limits and excellent customer service.

**Aetna International**

Aetna International provides comprehensive health insurance plans for expats, including coverage for inpatient and outpatient care, emergency evacuation, and wellness services. They have a strong global network and offer 24/7 support for their members.

### Allianz Worldwide Care

Allianz Worldwide Care offers robust international health insurance plans with comprehensive coverage, including inpatient and outpatient care, emergency medical evacuation, and dental and vision options. They are known for their high-quality service and comprehensive global network.

### GeoBlue

GeoBlue specialises in international health insurance, providing extensive coverage options tailored to expats. Their plans include inpatient and outpatient care, emergency evacuation, preventive care, and wellness programmes. GeoBlue is known for its user-friendly online services and strong customer support.

## Comparing Local and International Plans

Choosing the right health insurance plan is a critical decision for any expat in Thailand. With the variety of local and international options available, it can be challenging to determine which plan best suits your needs. In this section, we'll compare local and international insurance plans, examining their benefits and drawbacks to help you make an informed choice.

## Local health insurance plans

### Affordability

One of the primary advantages of local health insurance plans is their affordability. These plans are generally much cheaper than international alternatives, making them an attractive option for budget-conscious expats. Monthly premiums can range from THB 1,000 to THB 5,000, depending on the level of coverage and the insurance provider.

### Tailored to the Thai healthcare system

Local plans are designed specifically for the Thai healthcare system, ensuring smooth access to medical services nationwide. These plans typically have strong partnerships with local hospitals and clinics, making it easier to find in-network providers.

### Coverage limits and exclusions

While local plans are affordable, they often come with lower coverage limits and more exclusions than international plans. It's essential to read the fine print to understand what is and isn't covered. For example, some local plans may not cover pre-existing conditions, mental health services, or specialised treatments.

### Geographical limitations

Most local health insurance plans only provide coverage within Thailand. If you frequently travel outside the country, this could be a significant drawback. In such cases, you may need to purchase additional travel insurance to cover medical expenses abroad.

## International health insurance plans

### Comprehensive coverage

International health insurance plans are known for their comprehensive coverage, often including a wide range of medical services such as inpatient and outpatient care, emergency evacuation, dental and vision care, and mental health support. These plans are designed to cater to the needs of expats who require extensive and flexible coverage.

### Worldwide access

One of the most significant advantages of international plans is their worldwide coverage. Whether you're in Thailand, back home, or travelling to another country, you can access medical care without worrying about geographical limitations. This is particularly beneficial for digital nomads who frequently travel.

### Higher premiums

The extensive coverage and worldwide access of international plans come at a cost. Monthly premiums for international health insurance can range from THB 5,000 to THB 20,000 or more, depending on the level of coverage and the insurance provider. While this may be a significant investment, many expats find the peace of mind and flexibility worth the higher premiums.

### Customisable plans

International health insurance providers often offer customisable plans, allowing you to tailor your coverage to your specific needs. You can choose from various levels of coverage, add optional benefits such as dental and vision care, and adjust your deductible to manage your premiums.

### Customer service

Good customer service is crucial regarding health insurance, especially for expats facing language barriers and unfamiliar healthcare systems. Look for insurance providers that offer user-friendly apps and websites where you can easily make claims, check coverage details, and find in-network healthcare providers. An efficient online platform can save you time and hassle, making it easier to manage your health insurance.

Additionally, consider providers that offer 24/7 customer support. This ensures that you can get assistance whenever you need it, whether you're dealing with an emergency or just have a simple query. Reading reviews and asking for recommendations from fellow expats can give you insights into the quality of customer service offered by different providers.

**Conclusion**

Choosing between local and international health insurance plans requires careful consideration of your coverage needs, budget, and lifestyle. Local plans offer affordability and tailored coverage for the Thai healthcare system, while international plans provide comprehensive coverage and worldwide access. By understanding the key factors and comparing top providers, you can find the insurance plan that best suits your needs and ensures your health and well-being are taken care of. Stay covered and enjoy your expat journey with peace of mind!

# STAYING FIT & ACTIVE

Maintaining an active lifestyle is essential for your overall well-being, especially when living the digital nomad life in Thailand. Luckily, the Land of Smiles offers a plethora of fitness centres and gyms to help you stay in shape. From state-of-the-art facilities to community-driven spaces, there's something for everyone. Whether you prefer yoga, weightlifting, or group classes, Thailand has options to suit your fitness needs. Additionally, many gyms offer flexible memberships, making it easy to stay committed to your fitness goals. Let's explore the options and find the perfect place to break a sweat.

## Types of Fitness Centres and Gyms

### Commercial gyms

Commercial gyms are popular choices for many expats and locals alike. These gyms offer a wide range of equipment, classes, and amenities to cater to all fitness levels. Some well-known commercial gym chains in Thailand include Fitness First, Anytime Fitness, Virgin Active, and True Fitness.

**Pros**

- Extensive selection of equipment and facilities.

- Variety of group classes such as yoga, spinning, and high-intensity interval training (HIIT).
- Professional trainers and staff available for assistance and personal training.
- Convenient locations in major cities.

**Cons**

- Membership fees can be relatively high, ranging from THB 2,000 to THB 5,000 monthly.
- Can be crowded during peak hours.
- Limited access to outdoor activities, which can be a downside if you prefer exercising in natural settings.
- Contracts can be inflexible, often requiring long-term commitments or charging high cancellation fees.

**Boutique fitness studios**

Boutique fitness studios focus on specific types of workouts, such as yoga, Pilates, CrossFit, or dance. These studios offer a more personalised and community-oriented experience compared to commercial gyms.

**Pros**

- Specialised classes and expert instructors.
- Smaller class sizes for a more personalised experience.
- Friendly and supportive community atmosphere.
- Innovative and varied workout options that keep routines interesting.

**Cons**

- Higher cost per class or membership compared to commercial gyms.
- Limited equipment and facilities for general workouts.
- Offering limited or no facilities/services, e.g. changing rooms, showers, towels, water.

- Less flexibility in scheduling, as classes may be at fixed times.

### Hotel gyms

For those who prefer flexibility and convenience, hotel gyms can be a great option. Many hotels in Thailand offer well-equipped fitness centres that are accessible to non-guests for a fee. This is particularly useful if you travel frequently and need a temporary workout solution.

**Pros**

- Convenient locations, often in central areas.
- Access to additional hotel amenities such as pools and saunas.
- Flexible day-pass options.
- Ideal for travellers who need a consistent workout option without long-term commitments.

**Cons**

- Limited availability of equipment and classes.
- Can be more expensive than regular gym memberships.
- Don't offer 24/7 access, typically only between 9am and 9pm, which is not ideal for night owls or early birds.
- Often lack the community atmosphere and specialised classes found in dedicated fitness centres.

## Finding the right fitness solution for you

### Location and convenience

When choosing a gym, consider its proximity to your home or workplace. A conveniently located gym will make it easier to stick to your fitness routine. Use Google Maps or fitness apps to find gyms in your area and check their reviews.

### Facilities and equipment

Assess your fitness goals and ensure the gym you choose has the necessary equipment and facilities to meet them. Whether you're into weightlifting, cardio, or group classes, ensure the gym offers what you need. Visit the gym in person to get a feel for the environment and check the condition of the equipment.

### Membership options and pricing

Compare membership options and pricing for different gyms. Many gyms offer various membership plans, including monthly, quarterly, and annual options. Some gyms also provide flexible day passes or class packages, which can be ideal if you're not looking for a long-term commitment.

### Group classes and personal training

If you enjoy group workouts or need personalised guidance, look for gyms that offer a variety of classes and personal training services. Group classes can be a fun and motivating way to stay fit, while personal training provides tailored support to help you achieve your goals.

### Community and atmosphere

The atmosphere and community at a gym can significantly impact your motivation and overall experience. Look for gyms with a welcoming and supportive environment. Many expats find it helpful to join gyms with a mix of locals and foreigners, fostering a diverse and inclusive community.

## Top Fitness Centres and Gyms

We couldn't possibly list every fitness centre and gym in Thailand, so here are our top picks for both chain and independent workout places.

## Chains – top 10

If you prefer the consistency and reliability of well-known gym chains, Thailand has plenty to offer. These fitness centres provide many facilities, from state-of-the-art equipment to diverse group classes and professional personal training services. Whether you're looking for 24/7 access, luxurious amenities, or specialised classes, these chains have something to suit every fitness enthusiast. Let's explore some of the top gym chains (listed alphabetically) that can help you stay fit and healthy nationwide.

### Absolute You

Absolute You is a leading boutique fitness studio specialising in yoga, Pilates, and rhythm cycling. With multiple locations across Thailand, Absolute You offers a range of classes designed to help you stay fit and balanced. Membership options include class packages and monthly plans, with prices starting around THB 2,500 per month.

### Anytime Fitness

Anytime Fitness is a global gym chain with several locations across Thailand. Known for its 24/7 access, Anytime Fitness provides a convenient option for those who prefer working out at unconventional hours. The gym offers a range of equipment, personal training services, and a friendly community atmosphere. Membership fees start at around **THB 1,800 per month**.

### Carve House

Carve is a high-end fitness studio known for its unique blend of functional training and yoga. Located in Phuket and Bangkok, Carve offers a range of classes designed to enhance strength, flexibility, and overall fitness. Membership options include class packages and monthly plans, with prices starting around **THB 3,000 per month**.

**Fitness First**

Fitness First is a well-established gym chain with multiple locations across Thailand. They offer a wide range of equipment, group classes, and personal training services. Facilities often include saunas, swimming pools, and relaxation areas. Membership fees range from **THB 2,500 to THB 4,500 per month**.

**Fitwhey Gym**

Fitwhey Gym is a popular fitness chain known for its comprehensive range of weightlifting and strength training equipment. They also offer fitness classes and personal training services. With locations in Bangkok and other major cities, Fitwhey Gym provides a solid workout environment. Membership fees start at around **THB 2,000 per month**.

**Jetts 24 Hour Fitness**

Jetts 24 Hour Fitness offers the convenience of round-the-clock access, perfect for those with busy schedules. They provide a variety of equipment and fitness classes, including HIIT and functional training. With locations in major cities, Jetts offers flexible membership plans starting at around **THB 1,500 per month**.

**The Edge Boutique Gym**

The Edge Boutique Gym is a fitness chain with several locations in Thailand, known for its focus on functional training, HIIT, and personal coaching. They offer a variety of classes and advanced equipment to help you achieve your fitness goals. Membership fees start at around **THB 2,500 per month**, with options for class packages and monthly plans.

**True Fitness**

True Fitness is another popular fitness chain in Thailand, offering comprehensive facilities, including various workout equipment, group classes, and personal training services. With locations in major cities, True Fitness provides a comfortable and well-equipped environment for all your fitness needs. Membership fees typically range from **THB 2,000 to THB 4,000 per month**.

### Virgin Active

Virgin Active is another popular gym chain known for its high-quality facilities and extensive class offerings. With locations in Bangkok and other major cities, Virgin Active gyms feature state-of-the-art equipment, swimming pools, and wellness areas. Membership fees typically range from **THB 3,000 to THB 5,000 per month**.

### WeFitness Society

WeFitness Society offers a high-end fitness experience with multiple locations across Bangkok. They provide state-of-the-art equipment, a variety of group classes, and luxurious amenities such as spas and saunas. Membership fees range from **THB 3,000 to THB 6,000 per mont**h.

## Independents – top 20

### Bangkok

### BASE

BASE Bangkok is a boutique fitness studio specialising in high-intensity interval training (HIIT) and strength workouts. They offer small group classes and personal training in a supportive and community-focused environment. Class packages and memberships are available, starting at around **THB 2,500 per month**.

### F45 and FS8

F45 and FS8 are part of an international fitness franchise known for their functional training and team-based workout sessions.

F45 focuses on high-intensity interval training (HIIT), while FS8 combines elements of Pilates, yoga, and strength training. Both centres offer flexible membership options and class packages. Prices typically start around **THB 3,000 per month**.

**Siam Sports Pro**

Siam Sports Pro is one of Bangkok's premier tennis centres, offering top-quality courts and professional coaching. The facility caters to all skill levels, from beginners to advanced players, and provides group and private lessons. Membership options and court rental fees vary, starting around **THB 1,500 monthly**.

**The LAB**

The LAB is a boutique fitness studio in Bangkok offering a variety of functional training classes, including HIIT, strength training, and conditioning. With expert trainers and a friendly atmosphere, The LAB is an excellent choice for those looking to improve their fitness. Membership fees vary, with options for class packages and monthly memberships.

**Chiang Mai**

**CrossFit Chiang Mai / CFCNX**

CrossFit Chiang Mai, also known as CFCNX, is a top CrossFit affiliate in northern Thailand. The gym offers traditional CrossFit classes, Olympic lifting, and functional fitness training. With experienced coaches and a supportive community, CFCNX is a great place to achieve your fitness goals. Membership fees vary, with options for class packages and monthly memberships starting around **THB 3,000 per month**.

**GoGym**

GoGym is a popular fitness centre in Chiang Mai, offering a range of workout options, including free weights, cardio machines, and functional training areas. The gym also provides group classes like yoga, spinning, and circuit training. Personal training services are available for those seeking a customised fitness plan. Membership fees start at around **THB 1,500 per month**.

### Gym Bangarang

Gym Bangarang is a comprehensive fitness facility in Chiang Mai, known for its Muay Thai (thai boxing) training. They offer classes for all skill levels, from beginners to advanced fighters. In addition to Muay Thai, they provide strength and conditioning sessions, yoga, and personal training. Membership options include class packages and monthly plans, with prices starting around **THB 2,500 per month**.

## Hua Hin

### Sly Black Tennis Academy

Founded by former professional player and seasoned coach Sly Black, the Sly Black Tennis Academy is renowned for its top-tier tennis coaching and exceptional facilities. Catering to beginners and advanced players alike, the academy offers group and private lessons tailored to all skill levels. In addition to premier tennis courts, the facility features a fitness centre and various fitness classes to complement your training. Classes cost **THB 850-4,300 per session**.

### True Arena Hua Hin

True Arena Hua Hin is a state-of-the-art sports complex offering a wide range of fitness facilities, including a gym, tennis courts, and a swimming pool. They provide group fitness classes, personal training, and sports-specific training programmes. Membership fees vary, with options for daily, monthly, and annual passes starting around **THB 3,000 per month**.

**Koh Lanta**

**Lanta Gym Muay Thai**

Lanta Gym Muay Thai is a popular training facility in Koh Lanta that specialises in Muay Thai. They offer classes for all levels, from beginners to advanced practitioners. In addition to Muay Thai, the gym provides strength and conditioning sessions to enhance your fitness. Membership options include class packages and monthly plans, with prices starting around **THB 2,000 per month**.

**Lanta Sport Resort**

Lanta Sport Resort is a comprehensive fitness centre offering various training options, including Muay Thai, weightlifting, and cardio workouts. The facility also features an outdoor swimming pool and a sauna. They provide group classes and personal training services. Membership fees start at around **THB 2,000 per month**, with options for daily and weekly passes.

**Koh Samui**

**Elite Gym and Fitness**

Elite Gym and Fitness is a well-known gym on Koh Samui, offering a variety of fitness equipment, group classes, and personal training services. The gym features free weights, cardio machines, and functional training areas. They also offer Muay Thai and yoga classes. Membership options include daily passes and monthly plans, with prices starting around **THB 1,800 per month**.

**Superpro Samui**

Superpro Samui is a renowned training centre offering Muay Thai, MMA, and fitness classes. The facility includes a well-equipped gym, multiple training rings, and an outdoor pool. They cater to all skill levels, from beginners to professional fighters. Membership options include class packages and monthly plans, with prices starting around **THB 2,500 per month**.

### Krabi

#### Krabi Fitness

Krabi Fitness is a well-equipped gym in Krabi, offering a range of workout options, including free weights, cardio machines, and functional training areas. The gym also provides group classes such as yoga, Pilates, and circuit training. Personal training services are available for those looking for a customised fitness plan. Membership fees start at around **THB 1,500 per month**.

#### Krabi Lion Muay Thai

Located in the heart of Krabi, this is a premier training facility offering authentic Muay Thai training for all levels. Known for its experienced trainers, many of whom are former professional fighters, the camp provides tailored training programs through group classes and private sessions. Facilities include well-equipped training areas with heavy bags, pads, and a professional-grade boxing ring. Flexible membership options start at approximately **THB 2,000 per month**.

### Phuket

#### Alpha Health Club

Alpha Health Club is a well-equipped fitness centre in Phuket, offering a wide range of facilities, including free weights, cardio machines, functional training areas, and an outdoor pool. They also provide group classes such as yoga, spinning, and HIIT. Personal training services are available for those seeking a more tailored workout plan. Membership fees start at around **THB 2,000 per month**.

**Bangtao MuayThai and MMA**

Bangtao MuayThai and MMA is a premier training facility in Phuket offering classes in Muay Thai, Brazilian Jiu-Jitsu, and mixed martial arts (MMA). The gym caters to all levels, from beginners to professional fighters. They also provide strength and conditioning sessions to complement martial arts training. Class packages and memberships are available, with prices starting at around **THB 2,500 per month**.

**Delta 55 Fitness**

Delta 55 Fitness is a boutique gym in Phuket focusing on functional training and strength conditioning. They offer a variety of classes, including HIIT, kettlebell workouts, and mobility training. The gym prides itself on its community atmosphere and expert coaching staff. Membership options include class packages and monthly plans, with prices starting around **THB 2,500 per month**.

**Padel Phuket**

Padel Phuket, with locations in Blue Tree and Chalong, is the go-to destination for padel tennis enthusiasts in Phuket. The facility features modern courts and offers lessons for all levels, from beginners to advanced players. Padel Phuket also hosts tournaments and social events, creating a vibrant community atmosphere. Membership and court rental fees vary, starting around **THB 2,000 per month**.

**Thanyapura**

Thanyapura is a world-class sports and health resort in Phuket, renowned for its triathlon training facilities. The centre offers various fitness options, including swimming, cycling, and running tracks, as well as strength and conditioning training. Thanyapura provides expert coaching and state-of-the-art facilities, making it a top choice for triathletes and fitness enthusiasts. Membership and training packages vary, with prices starting around **THB 4,000 per month**.

## Yoga Studios, Wellness Centres, and Meditation Retreats

Thailand is a haven for those looking to maintain a balanced and active lifestyle. Beyond traditional gyms, the country offers a plethora of yoga studios, wellness retreats and outdoor activities that cater to various fitness levels and interests. Let's explore some of the best options to keep both your body and mind in top shape.

**Yoga studios**

Thailand offers a variety of yoga studios catering to all levels of practitioners. From bustling Bangkok to serene island retreats, these studios provide high-quality instruction in various styles, creating perfect spaces for relaxation, rejuvenation, and personal growth.

**Yoga Elements Studio (Bangkok)**

Located in the heart of Bangkok, Yoga Elements Studio is known for its high-quality instruction and diverse class offerings. From beginner to advanced levels, the studio provides classes in Vinyasa, Hatha and restorative yoga, among others. Drop-in rates are approximately **THB 500**, with packages and memberships available for regular attendees. The studio's central location and modern facilities make it a convenient and comfortable choice for urban yogis.

**The Yoga Tree (Chiang Mai)**

The Yoga Tree is a popular yoga studio in Chiang Mai, offering a variety of classes in a serene and welcoming environment. With experienced instructors and a range of styles, including Hatha, Vinyasa and Yin, this studio caters to all levels of practitioners. Drop-in classes are available for around **THB**

**300**, with membership options for frequent visitors. The calm garden setting adds to the overall experience, making it a perfect place to unwind and rejuvenate.

**Yoga Republic (Phuket)**

Yoga Republic in Phuket offers a wide range of classes, including Ashtanga, Vinyasa and Yin yoga. The studio is known for its skilled instructors and welcoming atmosphere, making it a great spot for both locals and tourists. Drop-in classes are around **THB 400**, with various package deals for those planning to stay longer. The studio also hosts workshops and special events, adding to its vibrant community feel.

**Lanta Yoga (Koh Lanta)**

Lanta Yoga on Koh Lanta provides a peaceful retreat with classes set in a beautiful natural environment. The studio offers a mix of Hatha, Vinyasa and restorative yoga, suitable for all levels. Drop-in rates are about **THB 300**, and they also offer multi-class packages. The serene island setting makes this studio an ideal place to deepen your practice and connect with nature.

**Yoga Balance (Krabi)**

Yoga Balance in Krabi is a great spot for those seeking a balanced mix of physical and mental well-being. The studio offers classes in various styles, including Hatha, Vinyasa and Yin yoga. Drop-in classes are priced at around **THB 350**, with discounts for multiple sessions. The friendly instructors and relaxed atmosphere make it a popular choice for both beginners and experienced practitioners.

**Wellness centres**

Thailand is renowned for its wellness retreats, offering everything from detox programmes to holistic healing. These retreats provide a comprehensive approach to health, combining fitness, nutrition and relaxation to help you reset and rejuvenate.

**Absolute Sanctuary (Koh Samui)**

Absolute Sanctuary is a well-known wellness retreat on Koh Samui, offering a range of programmes, including detox, yoga and fitness. The retreat features luxurious accommodations, healthy cuisine and a variety of wellness therapies. Programmes range from three days to several weeks, with prices starting around THB 20,000. Guests can enjoy amenities such as a saltwater swimming pool, steam room and a dedicated yoga studio. The retreat also provides nutritional consultations to help you maintain your wellness journey.

**Chiva-Som (Hua Hin)**

Chiva-Som is a luxury wellness resort in Hua Hin, known for its personalised wellness programmes and world-class facilities. The resort offers a range of treatments, including detox, weight management and stress relief. Packages are customised to meet individual needs, with prices starting from THB 50,000 for a three-day programme. The resort boasts state-of-the-art amenities, including a spa, fitness centre and hydrotherapy suite. Guests can also participate in activities like Tai Chi, Pilates and cooking classes.

**Kamalaya (Koh Samui)**

Kamalaya on Koh Samui is a holistic wellness sanctuary offering a range of programmes focused on detox, stress management and emotional balance. The retreat combines ancient healing traditions with modern wellness practices to create a transformative experience. Package prices start around THB 30,000. The retreat features beautiful accommodations, a world-class spa and a variety of wellness activities, including meditation and yoga classes. The on-site restaurant serves nutritious, gourmet meals designed to complement your wellness journey.

**Phuket Cleanse (Phuket)**

Phuket Cleanse is a wellness retreat in Phuket specialising in detox, fitness and holistic health. The retreat offers a range of programmes that include activities like yoga, Muay Thai, and HIIT. Prices for a week-long stay start around THB 40,000. The retreat also provides nutritional education and cooking classes to help you maintain a healthy lifestyle after your stay. Guests can enjoy amenities such as a sauna, ice bath and various wellness workshops.

**The Sanctuary (Koh Phangan)**

The Sanctuary in Koh Phangan is a holistic wellness retreat offering detox programmes, yoga courses and a range of therapeutic treatments. Nestled in a tranquil bay, the retreat provides a serene environment for relaxation and healing. Prices vary depending on the programme and length of stay, with detox packages starting at around THB 15,000. The Sanctuary also features a delicious vegetarian restaurant and offers workshops on topics like mindfulness and personal development.

**Meditation retreats**

For those looking to deepen their meditation practice and find inner peace, Thailand offers numerous retreats. These retreats provide a focused environment for self-discovery and growth, often set in picturesque locations. Whether you're a beginner or an experienced meditator, there's a retreat to suit your needs.

**Dipabhāvan Meditation Center (Koh Samui)**

Dipabhāvan Meditation Center is a serene retreat on Koh Samui, offering courses focusing on Vipassana meditation and mindfulness. The retreats typically last seven days and include daily meditation sessions, dhamma talks, and walking meditation. The cost is minimal, covering basic accommodation and meals. The stunning natural surroundings and experienced instructors make Dipabhāvan an excellent place to deepen your meditation practice.

**Suan Sati (Chiang Mai)**

Suan Sati is a meditation and yoga retreat centre near Chiang Mai. It offers programmes that combine traditional meditation practices with modern mindfulness techniques. The retreats include daily meditation and yoga sessions, as well as workshops on topics like breathwork and conscious living. Prices start around THB 15,000 for a seven-day retreat, including accommodation and meals. The centre's tranquil setting and holistic approach make it a great place to deepen your practice.

**The Mindfulness Project (Khon Kaen)**

The Mindfulness Project in Khon Kaen offers a variety of meditation retreats aimed at promoting mental clarity and emotional balance. The retreats combine mindfulness practices, yoga and community living. Prices for a week-long retreat start around THB 10,000, including accommodation and meals. The project's eco-friendly setting and supportive community provide an enriching environment for personal growth.

**Wat Pa Tam Wua Forest Monastery (Mae Hong Son)**

Wat Pa Tam Wua Forest Monastery offers a unique meditation experience in the heart of nature. The monastery welcomes visitors to join their meditation practice for free, though donations are appreciated. The serene forest setting, combined with the guidance of the resident monks, provides an ideal atmosphere for meditation and mindfulness. Accommodation and meals are basic but sufficient, allowing you to focus entirely on your practice.

### Wat Suan Mokkh (Surat Thani)

Wat Suan Mokkh is a Buddhist monastery in Surat Thani, offering ten-day silent meditation retreats. These retreats provide an immersive experience in mindfulness and meditation, guided by experienced monks. The cost of the retreat is minimal, typically around THB 2,000, covering basic accommodation and meals. The monastery's peaceful setting and structured schedule make it a perfect place for deep introspection.

## Conclusion

Staying fit and active in Thailand goes beyond traditional gyms and fitness centres. With many yoga studios, wellness retreats and outdoor activities, you'll find plenty of ways to maintain a healthy and balanced lifestyle. Whether you prefer the tranquillity of a daily yoga class, the rejuvenation of an occasional wellness retreat or the thrill of outdoor adventures, Thailand offers something for everyone. Embrace the opportunities provided by boutique fitness studios, meditation retreats and diverse sporting facilities. By exploring these options, you can ensure your body and mind stay in peak condition. Enjoy staying fit in the Land of Smiles!

### Stay tuned . . .

Get ready to immerse yourself in the rich tapestry of Thai culture in **Chapter 6: Cultural Insights and Learning Thai**. Understanding the local customs, traditions, and social norms is key to making your experience in Thailand more fulfilling and enjoyable. We'll cover essential cultural values such as *sanuk* (fun) and *mai pen rai* (no worries) and provide practical tips on Thai etiquette, including how to greet and show respect. With this knowledge, you'll be better equipped to navigate social interactions and build meaningful connections.

We'll also explore vibrant festivals and celebrations, giving you a glimpse into the heart of Thai life. Additionally, we'll arm you with basic Thai phrases to help you connect with locals and enhance your daily experiences. Stay tuned as we dive into Thailand's cultural landscape, making your transition smoother and your experience richer. Whether you're a seasoned traveller or new to the Land of Smiles, this chapter will offer valuable insights to deepen your understanding and appreciation of Thai culture.

# Chapter 6:
# Cultural Insights & Learning Thai

## Embracing Thai Culture: Traditions, Language and Social Norms

Welcome to the fascinating world of Thai culture! This chapter is your gateway to understanding the rich traditions and social norms that make Thailand so unique. Whether it's the concept of 'sanuk' (fun) or the relaxed 'mai pen rai' (no worries) attitude, you'll find that Thai culture is both welcoming and intricate. We'll explore essential etiquette, major festivals, and helpful tips for overcoming language barriers. Additionally, we'll guide you in learning the Thai language, enhancing your ability to connect with locals and enriching your overall experience. Get ready to immerse yourself in the Land of Smiles and navigate social interactions confidently and gracefully.

## UNDERSTANDING THAI CULTURE

Thailand's rich cultural tapestry is woven from centuries of traditions, values and philosophies that shape everyday life. As a digital nomad, getting to grips with these cultural nuances can not only enhance your experience but also help you integrate more smoothly into Thai society. In this section, we'll delve into key cultural values and traditions, as well as explore some quintessential Thai concepts.

### Key Cultural Values and Traditions

Thailand is a country rich in cultural heritage and traditions that have been passed down through generations. Understanding these key values will give you a deeper insight into what makes Thai society tick.

## Respect

One of the most important aspects of Thai culture is 'respect'. Respect is deeply ingrained in Thai society and can be seen in everyday interactions, from the way people greet each other with a 'wai' (a traditional Thai greeting involving a slight bow with the palms pressed together) to the way they speak and behave. Respect is not just reserved for elders but extends to teachers, monks, and even the royal family. The concept of 'saving face' is also crucial, meaning that Thais go to great lengths to avoid public embarrassment or confrontation.

## Family

In Thailand, family is the cornerstone of society, and strong family ties are highly valued. Extended families often live together or in close proximity, and it is common for multiple generations to share a household. Respect for elders is a significant part of this family-oriented culture. Family gatherings are frequent, and major decisions are often made collectively. As a digital nomad, understanding the importance of family can help you appreciate the societal structure and values.

## Religion

Religion plays a crucial role in the lives of many Thais, with the majority of the population practising Theravada Buddhism. Buddhist temples, or 'wats', are found throughout the country and are not just places of worship but also centres of community life. Monks are highly respected, and many young Thai men spend time in monastic life, even if only temporarily (i.e. one month). Religion influences many aspects of Thai life, from festivals and holidays to daily rituals and practices. As a digital nomad, visiting temples and understanding Buddhist customs can provide a richer cultural experience.

### Harmony

Thai culture places a high value on harmony and conflict avoidance. People tend to avoid confrontation and seek to maintain harmonious relationships. This is reflected in their communication style, which is often indirect and polite. Saving face, both one's own and others, is important, and criticism is usually delivered in a gentle and non-confrontational manner. This emphasis on harmony can be seen in various aspects of daily life, from business dealings to social interactions.

### Hospitality

Hospitality is a hallmark of Thai culture. Thais are known for their warmth and generosity towards guests, often going out of their way to make visitors feel welcome. This hospitality extends to the sharing of food, which is an integral part of social interactions. Meals are often communal, and sharing food is a way to strengthen bonds and create a sense of community. As a digital nomad, you will likely experience this hospitality firsthand, whether you're invited to a local's home or simply interacting with people at a market or restaurant.

## Quintessential Thai Concepts

Four quintessential concepts that epitomise Thai culture are 'sanuk', 'mai pen rai', 'sabai sabai', and 'kreng jai'. Additionally, understanding the deep respect and reverence for the Royal Family is crucial. These philosophies and cultural elements encapsulate the Thai approach to life and can greatly enhance your experience as a digital nomad in Thailand.

### Sanuk

'Sanuk' can be translated as 'fun', but it means much more than that. It is an essential part of the Thai way of life and reflects a belief that life should be enjoyable. Whether at work, socialising, or engaging in daily activities, Thais strive to make everything they do enjoyable and fun. This focus on enjoyment is evident in the lively markets, vibrant festivals* and the general joie de vivre that permeates Thai society. Embracing 'sanuk' means finding joy in the little things and approaching life with a light-hearted attitude.

- **Work:** Even at work, Thais try to incorporate an element of fun. Offices often have a relaxed atmosphere and colleagues frequently engage in light-hearted banter.
- **Social gatherings:** Thai social gatherings are characterised by music, dance, and laughter. Whether it's a family gathering or a public festival, 'sanuk' is always a central element.
- **Daily activities:** From shopping at bustling markets to dining at street food stalls, daily activities are infused with a sense of fun and enjoyment.

**Mai pen rai**

'Mai pen rai' translates to 'no worries' or 'it's okay'. This phrase captures the Thai attitude of not sweating the small stuff and maintaining a laid-back, relaxed approach to life. It's a way of expressing that things will work out in the end and there's no need to stress. This mindset can be particularly refreshing for those coming from fast-paced, high-pressure environments. Adopting a 'mai pen rai' attitude can help you better navigate the inevitable challenges and uncertainties that come with being a digital nomad, allowing you to remain calm and adaptable.

**Problem-solving:** Thais often approach problems with a 'mai pen rai' attitude, focusing on finding a solution rather than dwelling on the issue.

- **Interpersonal relationships:** This philosophy helps in maintaining harmonious relationships, as it encourages letting go of grudges and moving forward.

- **Daily life:** Whether it's dealing with traffic jams or unexpected delays, 'mai pen rai' helps maintain a calm and positive outlook.

### Sabai sabai

'Sabai sabai' translates to 'take it easy' and embodies the Thai ethos of relaxation and comfort. This concept encourages a balanced lifestyle where stress is minimised, and well-being is prioritised. It's about enjoying life at a leisurely pace and not letting the pressures of daily life overwhelm you. Embracing 'sabai sabai' can help you adapt to the slower, more relaxed pace of life in Thailand, making your experience as a digital nomad more enjoyable and fulfilling.

- **Leisure time:** Thais value downtime and relaxation. Whether it's spending time at a spa, enjoying a leisurely meal, or simply lounging at home, 'sabai sabai' is about taking the time to relax and recharge.
- **Work-life balance:** In the workplace, 'sabai sabai' promotes a balanced approach to work. While tasks are completed efficiently, there's an emphasis on not letting work-related stress take over your life.
- **Keep it cool:** Daily interactions are approached with a calm and easy-going attitude. Whether you're dealing with a minor inconvenience or a more significant issue, 'sabai sabai' encourages you to stay calm and not let it affect your peace of mind.

### Kreng jai

'Kreng jai' is a uniquely Thai concept that translates to 'consideration' or 'deference'. It involves being mindful of others' feelings and avoiding actions that might inconvenience or burden them. This cultural value is deeply ingrained in Thai society and influences social interactions and behaviours.

Embracing 'kreng jai' can help you navigate social situations with grace and sensitivity.

**Avoidance:** Thais often go out of their way to avoid causing inconvenience or discomfort to others. This might mean not expressing negative emotions openly or declining an offer if it seems burdensome to the person making it.

- **Deferential behaviour:** Thais show respect and consideration through their actions. This can include speaking softly, avoiding direct confrontation, and being mindful of others' needs and feelings.
- **Social harmony:** 'Kreng jai' helps maintain social harmony by encouraging individuals to be considerate and respectful in their interactions. This creates a more pleasant and cooperative environment for everyone.

## The Thai Royal Family

The Royal Family holds a revered place in Thai society, and their influence extends into many aspects of daily life. The monarchy is deeply respected, and the King is often seen as a unifying figure for the country. Understanding the importance of the Royal Family and showing respect for them in conversations and actions is crucial. This respect is reflected in various aspects of daily life, from standing still for the national anthem to avoiding any negative remarks about the monarchy. Learning about the Royal Family and their role in Thai culture can help you navigate social interactions more gracefully and respectfully.

### The cornerstone of Thai society

There are six crucial areas to comprehend and adhere to in the Land of Smiles if you don’t want to cause offence, be embarrassed or even end up in jail.

**National unity**

The Royal Family, particularly the King, is seen as a symbol of national unity and stability. Their presence and actions often serve to bring the country together, especially during times of crisis. The King is viewed as a father figure and a moral compass for the nation, guiding and supporting the people. The Queen and other members of the Royal Family also play significant roles in various social and cultural initiatives, contributing to the country's well-being.

**Public displays of respect**

Thais show their respect for the Royal Family in various ways. One of the most noticeable is standing for the national anthem, which is played in public places twice a day and before film/theatre shows. It's a moment of unity and respect that everyone participates in. Additionally, portraits of the King and Queen are commonly displayed in homes, businesses, and public spaces. These portraits are often adorned with flowers and other decorations, symbolising reverence and loyalty.

**Cultural events**

Many cultural events and public holidays are centred around the Royal Family, celebrating birthdays, anniversaries, and other significant milestones. For instance, the King's birthday on 28th July and the Queen's birthday on 12th August are national holidays, celebrated with grand ceremonies, parades, and public festivities. These events offer a unique opportunity to witness the deep connection between the monarchy and the people, providing deeper insights into Thai traditions and values.

**Contributions to society**

The Royal Family is involved in various charitable and developmental projects across Thailand. These initiatives encompass various areas, including education, healthcare, agriculture, and environmental conservation. The late King

Bhumibol Adulyadej, in particular, was known for his numerous royal projects aimed at improving the quality of life for the Thai people. His dedication to sustainable development and self-sufficiency has left a lasting legacy that inspires the nation.

**The King's role in Buddhism**

The King of Thailand is also considered the upholder of Buddhism, the country's predominant religion. The Royal Family's involvement in religious ceremonies and their patronage of Buddhist temples underscore their role in preserving and promoting the cultural and spiritual heritage of Thailand. This connection reinforces the moral and ethical values that the monarchy embodies, further strengthening their bond with the people.

**Avoiding negative remarks**

It's essential to understand that the Royal Family is highly revered, and any negative remarks or criticism about the monarchy are not only socially unacceptable but also illegal under Thailand's strict lèse-majesté laws. These laws protect the dignity of the Royal Family, and violations can result in severe penalties. As a digital nomad, it's crucial to be aware of these sensitivities and approach conversations about the monarchy respectfully and cautiously.

## Conclusion

Understanding Thai culture is your key to unlocking a more enriching and harmonious experience in the Land of Smiles. By appreciating the core values of respect, family, religion, harmony, and hospitality, you'll find yourself better equipped to integrate into Thai society. Embracing the concepts of 'sanuk', 'mai pen rai', 'sabai sabai', and 'kreng jai' will not only help you adapt to the local lifestyle but also bring a refreshing perspective to your own life. Participating in nationwide and local festivals and celebrations will deepen your cultural appreciation. So go ahead, dive in, and enjoy the vibrant, welcoming culture that Thailand has to offer.

## THAI ETIQUETTE

Thai etiquette is a fascinating blend of tradition, respect, and social harmony. As a digital nomad, understanding the dos and don’ts of Thai etiquette will help you avoid faux pas and enrich your interactions with locals. In this section, we'll guide you through common dos and don'ts and the proper ways to greet and show respect. Embrace these cultural nuances, and you'll find yourself navigating Thai society easily and gracefully. From the traditional wai greeting to showing deference to elders, mastering these customs will make your experience in Thailand more rewarding and respectful.

### Common Dos and Don'ts

Thailand's social etiquette is deeply rooted in respect and harmony. Here are some essential dos and don'ts to keep in mind:

**Dos**

- **Use the 'wai':** The 'wai' is a traditional Thai greeting involving a slight bow with the palms pressed together in a prayer-like gesture. It's a sign of respect and is used in various social interactions.
- **Dress modestly:** When visiting temples or religious sites, dress conservatively. Shoulders and knees should be covered. This shows respect for the cultural and religious significance of these places.
- **Remove your shoes:** Always remove your shoes before entering someone's home or a temple. This is a sign of respect and cleanliness.

- **Be polite and soft-spoken**: Thais value politeness and a calm demeanour. Speak softly and avoid raising your voice, as it can be seen as aggressive or disrespectful.
- **Respect elders:** Show deference to older individuals by allowing them to go first, offering them seats, and addressing them with respectful titles.
- **Use both hands when giving or receiving:** When handing over or receiving items, especially from elders, use both hands as a sign of respect.

**Don'ts**

- **Don't touch someone's head:** The head is considered the most sacred part of the body, so avoid touching anyone's head, even in a friendly manner.
- **Avoid pointing your feet:** Feet are considered the lowest and dirtiest part of the body. Don't point your feet at people or religious objects and avoid resting your feet on tables or chairs.
- **Don't lose your temper:** Displaying anger or frustration in public is frowned upon. Thais value 'saving face', so remain calm even in challenging situations.
- **Avoid public displays of affection:** While holding hands is generally acceptable, more intimate displays of affection, such as kissing, are considered inappropriate in public.
- **Don't touch monks:** Women should avoid touching monks or handing them items directly. If a woman needs to give something to a monk, it's best to place it on a cloth or a table for the monk to pick up.
- **Avoid pointing with your finger:** Instead of pointing with a finger, use your entire hand to gesture. Pointing is considered rude and aggressive.

## Traditional Thai Etiquette

The 'wai' is the cornerstone of Thai greetings and a fundamental way to show respect. Here's how to do it correctly and understand its nuances:

### Performing the 'wai'

- **Hand positioning:** Place your palms together in a prayer-like gesture at chest level. The higher the hands, the more respect you are showing. However, for most social interactions, chest level is appropriate.
- **Bowing:** Slightly bow your head while keeping your palms together. The depth of the bow can vary depending on the level of respect you wish to convey.
- **Returning the 'wai':** If someone offers you a 'wai', it's polite to return it. However, exceptions exist, such as when receiving a 'wai' from a child or someone in a service role (like a waiter).
- **Knowing when to initiate:** Generally, the person of lower status or younger age initiates the 'wai'. However, it's always safe to reciprocate a 'wai' when offered.
- **Avoiding the 'wai' in certain contexts**: While the 'wai' is a sign of respect, it is not typically used in casual contexts such as with close friends. A simple nod or smile can suffice.

### Respecting elders and superiors

**Use formal language:** Titles and polite particles like 'khun' (Mr/Ms) followed by the person's first name show respect. Adding 'ka' (for women) or 'krub' (for men) at the end of sentences also adds politeness.

- **Offer your seat:** If you're on public transport and see an elderly person, offer them your seat. This simple act of kindness is highly appreciated.

- **Lower yourself:** When passing by someone seated, especially an elder or monk, try to lower your body slightly as a sign of respect.
- **Avoid interrupting:** When an elder or superior is speaking, avoid interrupting them. Wait for them to finish before you respond.
- **Use both hands:** When handing over or receiving items from an elder or superior, use both hands to show respect and politeness.

**Everyday behaviour**

Thailand is a country where religion and spirituality play a central role in daily life. Temples, or 'wats', are not just places of worship but also hubs of community activity. Understanding how to behave in these sacred spaces is crucial for respecting the local culture.

**Dress appropriately**

As mentioned earlier, dress modestly when visiting temples. This typically means covering your shoulders and knees. In some temples, you may be required to wear long pants or skirts, so it's a good idea to carry a sarong or shawl with you. Some places may have them available to borrow or rent.

**Behave respectfully**

Keep your voice low, don't point at religious statues, and always walk around rather than in front of someone praying. When sitting, avoid pointing your feet towards Buddha statues or monks. Instead, sit so your feet are tucked to the side or behind you.

**Respect religious rituals**

When witnessing or participating in religious rituals, maintain a respectful distance and follow any instructions from locals or temple staff. Avoid interrupting ceremonies and always ask for permission before taking photos. If you're unsure about something, observe the locals or ask politely.

### Making merit

Making merit, or 'tam boon', is a common practice in Thailand. This involves activities such as offering food to monks, donating to temples, or releasing animals like fish or birds as a symbolic act of kindness. Participating in these activities can be a meaningful way to engage with local customs.

### Offerings

When making offerings, whether food, flowers, money, or incense, do so with both hands as a sign of respect. If you're unsure about the proper procedure, watch the locals or ask for guidance. Remember to make the offering with an open heart and a smile.

## Additional etiquette tips

### Dining manners

When eating with locals, wait for the eldest person to start eating before you begin. Sharing food is common, and it's polite to try a bit of everything. If you're eating with a group, avoid taking the last piece of food on a shared plate without offering it to others first.

### Gift-giving

If you're invited to someone's home, bringing a small gift such as fruit, flowers, or sweets is a nice gesture. Avoid giving gifts with sharp objects, like knives, as they can symbolise severing a relationship.

### Public behaviour

Loud and boisterous behaviour is generally frowned upon. Thais appreciate a calm and composed demeanour, especially in public spaces. Always use your right hand when giving or receiving items, as the left hand is considered unclean.

**Interactions with locals**

When interacting with locals, maintaining a polite and respectful tone is crucial. Avoid criticising or questioning someone too directly, as this can cause them to lose face. Instead, approach sensitive topics with tact and diplomacy.

**Showing gratitude**

Expressing gratitude is important in Thai culture. A simple 'thank you' ('khop khun') or a small gesture of appreciation, such as a smile or a nod, goes a long way. When receiving gifts or favours, show your appreciation sincerely.

**Conclusion**

Mastering Thai etiquette is your passport to smooth and respectful interactions in Thailand. By following common dos and don'ts and learning how to properly greet and show respect, you'll find yourself warmly welcomed by locals. These small gestures of respect and understanding can go a long way in building meaningful connections and making your time in Thailand enjoyable and enriching. So go ahead, practise your 'wai', mind your manners and immerse yourself in the warm and respectful Thai culture. Whether visiting temples, joining in local celebrations, or simply sharing a meal, these insights will help you navigate Thai society with grace and confidence.

## FESTIVALS & CELEBRATIONS

Thailand's festivals and celebrations are a vibrant and integral part of its cultural landscape. These events are not just about fun and revelry; they also offer a window into the country's rich traditions, religious beliefs, and communal spirit. As a digital nomad, participating in these celebrations can provide a deeper understanding of Thai culture and create unforgettable memories. In this section, we'll explore major Thai festivals like Songkran and Loy Krathong, as well as other local holidays and how they are celebrated.

## The BIG Three

Thailand's festivals and celebrations are a vibrant and integral part of its cultural landscape. These events are not just about fun and revelry; they also offer a window into the country's rich traditions, religious beliefs, and communal spirit. As a digital nomad, participating in these celebrations can provide a deeper understanding of Thai culture and create unforgettable memories. In this section, we'll explore major Thai festivals like Songkran and Loy Krathong, as well as other local holidays and how they are celebrated.

### Songkran

Songkran, the Thai New Year, is one of the most eagerly anticipated festivals in Thailand. Traditionally held from the 13th to the 15th of April, this festival marks the beginning of the Thai solar year. Songkran is celebrated with water, symbolising the washing away of sins and bad luck. During Songkran, you'll find streets filled with people engaging in playful water fights, using everything from water guns to buckets. It's a time of joy, laughter, and community spirit. In addition to the water festivities, Thais also visit temples to make merit, offer food to monks, and pour water over Buddha statues as a sign of respect. It's a fantastic opportunity for digital nomads to join in the fun and experience the warmth of Thai hospitality.

#### Tips for enjoying Songkran

- Dress appropriately: Wear light, quick-drying clothes. Avoid white or see-through materials.
- Protect your belongings: Use waterproof bags for your phone, wallet, and other valuables.

- Be respectful: Remember that while it's fun, Songkran has deep cultural and religious significance. Be mindful when splashing water on elders or monks.
- Use non-permanent colours: If you choose to join in with the coloured powders, make sure they are non-toxic and washable.
- Stay hydrated: The weather can be hot, so drink plenty of water and take breaks in the shade.
- Join local traditions: Participate in traditional practices like pouring water over Buddha statues and visiting temples to make merit. This adds a spiritual dimension to the festive fun.

**Loy Krathong**

Loy Krathong, also known as the Festival of Lights, is celebrated on the full moon night of the 12th lunar month, usually in November. This festival is a beautiful and serene event where people gather around rivers, lakes, and canals to float 'krathongs' – small, elaborately decorated baskets made from banana leaves, flowers, and candles. The act of floating a krathong symbolises letting go of grudges, anger, and negative thoughts. It's a time for reflection and renewal. The sight of thousands of glowing krathongs drifting along the water is breathtakingly beautiful and offers a peaceful contrast to the lively Songkran festival. Many places also release lanterns into the sky, adding to the magical atmosphere. Participating in Loy Krathong is a wonderful way to connect with Thai culture and experience its more contemplative side.

**Tips for enjoying Loy Krathong**

Make your own krathong: Many locals make their own krathongs, which adds a personal touch to the celebration. You can join workshops or buy materials at local markets.

- Choose an eco-friendly krathong: Opt for biodegradable materials to minimise environmental impact.
- Find a good spot: Popular spots can get crowded. Arrive early to find a good viewing and launching spot.
- Respect local customs: Follow the locals' lead regarding when and how to launch your krathong.
- Capture the moment: While photos are great, take time to soak in the atmosphere and enjoy the festival's beauty.
- Participate in lantern releases: In some areas, like Chiang Mai, you can also release lanterns into the sky, adding to the enchanting ambience.

**Chinese New Year**

Chinese New Year is widely celebrated in Thailand, especially in areas with large Chinese communities, such as Bangkok's Chinatown. This vibrant festival is marked by dragon and lion dances, fireworks, and the exchange of red envelopes (hongbao) containing money. It's a time for families to come together, make offerings at temples, and celebrate with feasts. The colourful celebrations bring a festive atmosphere to the streets of Thailand, making it a unique experience for visitors.

**Tips for enjoying Chinese New Year**

- Visit Chinatown: Bangkok's Chinatown is the epicentre of Chinese New Year celebrations. The streets are adorned with red lanterns, and various performances and parades occur.
- Join the festivities: Participate in dragon and lion dance performances, which are believed to bring good luck and drive away evil spirits.
- Try traditional foods: Enjoy a variety of Chinese delicacies such as dumplings, spring rolls, and sweet rice cakes. Many restaurants offer special menus during the festival.

- Give and receive hongbao: The red money envelopes symbolise good luck and prosperity. If you're invited to join a family's celebration, it's customary to give hongbao.
- Make offerings: Visit local temples to make offerings and pray for good fortune in the coming year. This is a great way to immerse yourself in the spiritual aspect of the festival.
- Wear red: Red is the colour of good fortune and joy in Chinese culture. Wearing red during the celebrations is considered auspicious.

## Other Local Festivals

### Top 10

Thailand is renowned for its rich and diverse cultural events. Here are the top 10 local festivals (listed alphabetically) you should know about:

#### Full Moon Parties (Koh Phangan)

The Full Moon Party on Koh Phangan is one of the most famous beach parties in the world. Held monthly during the full moon, this vibrant event attracts thousands of revellers from around the globe. With abundant music, dancing, and fire shows, the party stretches along Haad Rin Beach and offers an unforgettable nightlife experience under the glowing full moon. Make sure to plan your accommodation and travel in advance, as the island becomes extremely busy during this time. Prices for attending Full Moon Parties can vary, but you can generally expect to spend **THB 500 to THB 1,000** for entrance, with additional costs for food, drinks, and accommodation.

#### Ghost Festival (Phi Ta Khon)

The Ghost Festival, or Phi Ta Khon, is held in the Dan Sai district of Loei Province. This lively and colourful event features participants dressed in ghostly masks and costumes, parading through the streets to the sound of music and dance. The festival is part of a larger event called Bun Luang, which includes religious ceremonies and merit-making activities. It's a celebration of local folklore and a fun, vibrant festival that showcases the unique cultural heritage of the region.

**Loi Ruea Chao Le Festival**

The Loi Ruea Chao Le Festival, also known as the Sea Gypsy Boat Floating Festival, is celebrated by the Moken and Urak Lawoi sea gypsy communities in southern Thailand. Held in April and October, this festival involves the building and floating of intricately decorated boats to dispel evil spirits and bring good luck. The event includes traditional music, dance, and rituals, providing a unique cultural experience.

**Makha Bucha Day**

Makha Bucha Day, observed on the full moon day of the third lunar month, commemorates the spontaneous gathering of 1,250 monks to hear Buddha's sermon. Thais celebrate this day by visiting temples to make merit, listen to sermons, and participate in candlelight processions. It's a day of spiritual reflection and a great opportunity to learn about Buddhist practices.

**Note on alcohol sales**: Alcohol is not sold on Thai Buddhist holidays, including Makha Bucha Day and Visakha Bucha Day. The only exceptions are hotel bars and some restaurants. It’s worth noting that certain local establishments might serve alcohol in disguise, such as in a generic cup, but generally, it won't be widely available. It's advisable to plan ahead if you wish to consume alcohol, as the restrictions are strictly enforced. Respecting these rules is part of honouring the local culture and traditions.

## Phuket Vegetarian Festival

The Phuket Vegetarian Festival is a nine-day event held in October. It's known for its extreme rituals, where participants perform acts of self-mortification, such as piercing their bodies with sharp objects, walking on hot coals, and climbing ladders made of blades. The festival is held to purify the participants and the community, and it includes processions, temple ceremonies, and, of course, a wide variety of vegetarian food. It's a fascinating and intense festival that offers a deep dive into Thai-Chinese culture.

## Rocket Festival (Bun Bang Fai)

The Rocket Festival is a traditional event held in the northeastern region of Thailand, particularly in Isan. It involves the launching of homemade rockets to encourage the gods to bring rain for the upcoming rice-growing season. The festival includes parades, music, dance, and the thrilling spectacle of rockets being launched into the sky. It's a lively and energetic celebration rooted in agricultural traditions.

## Surin Elephant Round-up

The Surin Elephant Round-up is an annual festival held in November in Surin Province. This event showcases the skills and strength of elephants in various performances, including parades, tug-of-war, and historical re-enactments. The festival highlights the importance of elephants in Thai culture and offers visitors a chance to see these magnificent animals up close.

## Thai National Day

Thai National Day, celebrated on the 5th of December, marks the birthday of the late King Bhumibol Adulyadej. It's a day of national pride, with various festivities, including parades, cultural performances, and fireworks. It's an excellent opportunity for digital nomads to join locals in celebrating Thai heritage and unity.

**Visakha Bucha Day**

Visakha Bucha Day, celebrated on the full moon day of the sixth lunar month, marks the birth, enlightenment, and death of Buddha. Like Makha Bucha Day, people visit temples to make merit and join candlelight processions. It's one of the most important Buddhist holidays and offers a deep insight into the spiritual side of Thai culture.

**Yi Peng Lantern Festival**

Yi Peng, often celebrated alongside Loy Krathong, is famous for its sky lanterns. Held primarily in Chiang Mai, this festival sees thousands of lanterns released into the night sky, creating a stunning visual spectacle. The lanterns, made from rice paper, are released with prayers and wishes for the future. This festival is a beautiful blend of light and spirituality, offering a unique experience for visitors.

**Finding event info**

Staying informed about upcoming events and festivals in Thailand can greatly enhance your experience. Here are some websites and resources to help you stay updated:

- **Bangkok Post**: The Bangkok Post frequently publishes articles about major events and festivals happening across Thailand, keeping you up to date.
- **BK Magazine**: A popular source for lifestyle news and events in Bangkok, offering insights into the city's vibrant social calendar.
- **Eventbrite**: This widely-used event management and ticketing website often lists local events and festivals happening throughout Thailand.
- **Lifestyle and Travel**: A magazine that covers travel, lifestyle, and events, providing detailed information on activities in Thailand.

- **Local Facebook Groups**: Joining local Facebook groups dedicated to events and activities in specific cities or regions can be very helpful.
- **Phuket.Net:** This site has comprehensive travel information about Phuket and its surroundings for tourists. Plus, there is comprehensive information about living in Phuket for expats.
- **Thailand Convention & Exhibition Bureau**: Provides information on business events, conventions, and exhibitions in Thailand.
- **Thailand Event Guide**: Offers a comprehensive list of events and festivals around the country, including details and dates.
- **Thailand Magazine**: Features articles on various events and festivals, making it easy to find something interesting to attend.
- **TAT Newsroom**: Offers the latest news on tourism and events in Thailand, ensuring you stay well-informed.
- **Time Out Magazine**: Provides detailed listings and reviews of events, concerts, and festivals in major cities across Thailand.
- **Tourism Authority of Thailand**: The official TAT website provides comprehensive information on events and festivals nationwide.
- **Thai Visa Forum**: This online forum is a valuable resource for expats and travellers in Thailand, often featuring discussions about upcoming events and local happenings.

By using these resources, you can ensure that you don't miss out on any of the exciting events and festivals that Thailand has to offer.

### Conclusion

Thailand's festivals and celebrations offer a colourful and immersive way to experience the country's rich cultural heritage. From the joyous water fights of Songkran to the serene beauty of Loy Krathong, each festival provides unique insights into Thai traditions and values. Local holidays, whether religious or national, further enrich your understanding of Thailand's communal and spiritual life. As a digital nomad, participating in these events will not only deepen your cultural appreciation but also create cherished memories. So, get ready to join the festivities, embrace the local customs, and enjoy the vibrant celebrations Thailand has to offer.

# OVERCOMING LANGUAGE BARRIERS

Language can be one of the most significant barriers when moving to a new country, but it can also be one of the most rewarding aspects to overcome. While English is widely spoken in tourist areas and big cities in Thailand, knowing some basic Thai phrases and effective communication strategies can go a long way in enhancing your experience. This section will equip you with essential Thai phrases and tips for effective communication, helping you navigate daily interactions more smoothly and build stronger connections with locals.

## Starting to communicate

### Tips for effective communication

Navigating a new language can be challenging, but with the right approach, you can communicate effectively and respectfully in Thailand.

#### Use simple language

When speaking English, use simple words and short sentences. Avoid slang and idiomatic expressions that might not translate well. Remember not to raise your voice, as Thais are averse to anything that seems confrontational.

**Body language**

Non-verbal cues are crucial. Smile often, as it's a universal sign of friendliness and can ease communication. Thais are known for their warm smiles, which can help bridge language gaps. Use hand gestures when appropriate.

**Speak slowly and clearly**

This helps in ensuring that your words are understood correctly. Avoid speaking louder, as it can come off as aggressive. Instead, focus on clear enunciation and a calm tone. At the same time, don't 'dumb down' your speech to a toddler level, as that can be condescending.

**Use translation apps**

Apps like **Google Translate** and **ChatGPT** can be lifesavers. They offer instant translations and can help bridge the gap in understanding. Some apps even have voice translation features, making real-time conversations easier.

**Learn the Thai script**

Familiarising yourself with the Thai script can be beneficial, especially when reading signs and menus. Even basic recognition of numbers and common words can be helpful. This can also impress locals and show your dedication to understanding their culture.

**Be patient and polite**

Patience goes a long way. If you do not understand the first time, try rephrasing or using gestures. Always maintain a polite tone. Thais appreciate calm and respectful communication, especially with a language barrier.

**Practice active listening**

Show that you're engaged in the conversation. Nod, make eye contact, and respond appropriately. This not only helps in understanding but also shows respect. Active listening can also help you pick up on context clues and improve your language skills.

**Cultural sensitivity**

Be aware of cultural nuances. For example, avoid touching someone's head or pointing your feet at people, as these actions are considered disrespectful in Thai culture. Understanding these cultural aspects can prevent misunderstandings and show your respect for Thai customs.

**Learn from locals**

Don't hesitate to ask locals for help with pronunciation or understanding. Most people appreciate the effort and are willing to assist. Engaging in these interactions can also provide valuable learning experiences and deepen your cultural immersion.

**Join language exchange groups**

These groups provide an excellent opportunity to practice Thai and help locals with English. It's a win-win situation and a great way to make friends. Look for local meetups, online forums, or social media groups dedicated to language exchange.

**Use visual aids**

When words fail, visual aids can be incredibly helpful. Carry a small notebook to draw simple pictures or write down important phrases. Pictures of common items, maps, or gestures can clarify your message when language barriers seem insurmountable.

**Practice with language apps**

Apart from translation apps, there are numerous language learning apps like **Duolingo, Mondly**, or **Pimsleur** that offer structured lessons in Thai. These can be a great way to build your vocabulary and improve pronunciation at your own pace.

**Engage in daily practice**

Consistent practice is key to language learning. Try to use Thai in your daily interactions, whether it's ordering food, asking for directions, or chatting with locals. The more you practice, the more confident and fluent you'll become.

**Why learning basic Thai is a good idea**

Learning basic Thai can greatly enrich your experience in Thailand. Here are some key reasons why it's worth the effort:

**Enhanced daily interactions**

While many Thais in urban areas and tourist spots speak some English, knowing basic Thai phrases can make everyday transactions – like ordering food, asking for directions, or shopping – much easier and more pleasant. You'll find that locals are often delighted and more willing to assist when they see you trying to speak their language.

**Cultural appreciation**

Language is a key component of culture. By learning Thai, you gain insights into Thai customs, traditions, and values. It allows you to understand nuances and subtleties often lost in translation. You'll be able to appreciate Thai literature, music, and media on a deeper level, enriching your cultural experience.

**Building relationships**

Speaking Thai can help you form stronger connections with locals. Whether with your neighbours, colleagues, or friends you meet along the way, a shared language fosters a sense of camaraderie and mutual respect. It opens doors to more meaningful conversations and deeper friendships.

**Professional advantages**

For digital nomads who plan to stay in Thailand for an extended period, knowing Thai can be a significant professional asset. It can facilitate networking, enhance business dealings, and provide opportunities for collaboration with local businesses and professionals.

**Safety and convenience**

In emergencies or tricky situations, knowing some Thai can be incredibly useful. Whether you need to communicate with healthcare providers, police, or other authorities, being able to speak and understand Thai can make a big difference.

**Respect and appreciation from locals**

Making an effort to learn and use the Thai language shows respect for the local culture and people. Thais greatly appreciate when foreigners try to speak their language, even if it's just a few basic phrases. This effort often garners gratitude and a warm response, demonstrating humility and a willingness to integrate into the community. By showing respect through language, you build goodwill and create a more welcoming environment for yourself.

## Conclusion

Mastering basic Thai can significantly enhance your experience in Thailand. While English is widely spoken in tourist areas, knowing some Thai phrases will make daily interactions smoother and more rewarding. Using simple language, practising active listening, and being culturally sensitive will help you build stronger connections with locals. Your efforts will be appreciated, as will showing respect for the local culture and fostering goodwill. Embrace the challenge, practice regularly, and engage with locals to make your stay in Thailand more immersive and enjoyable. Welcome to Thailand, and enjoy the journey of cultural and linguistic discovery!

# LEARNING THAI

Learning the Thai language can be both a challenging and rewarding experience. As a digital nomad, understanding some basic Thai phrases and communication strategies will greatly enhance your daily interactions and help you build stronger connections with locals. This section will guide you on where and how to study Thai effectively, providing you with the tools you need to immerse yourself in the language and culture. From language schools and private tutors to online courses and immersive experiences, you'll find a variety of methods to suit your learning style and pace.

## Mastering the Thai Language

### Where and how to study Thai

Learning Thai can be a rewarding experience that enhances your stay in Thailand. Whether you prefer structured classes or self-study methods, numerous options are available to help you master the language.

### Language schools

Enrolling in a language school is a structured and effective way to learn Thai. Schools like **ALA Language Center, Duke Language School**, and **Thai Language Hut** offer courses tailored to different proficiency levels. These schools provide a comprehensive curriculum, experienced teachers, and a supportive learning environment.

### Private tutors

Hiring a private tutor offers personalised instruction tailored to your learning pace and goals. Tutors can focus on areas you want to improve and provide immediate feedback. Websites like

**Italki** and **Preply** can help you find qualified Thai tutors for online or in-person lessons. You can also find experienced tutors via local **WhatsApp** and **Facebook Groups**.

**Community classes**

Many community centres and cultural organisations offer Thai language classes at affordable rates. These classes are often more relaxed and provide an opportunity to meet other learners and engage in cultural activities.

**Online courses**

If you prefer flexibility, online courses are a great option. Platforms like **Coursera, Udemy**, and **ThaiPod101** offer comprehensive Thai language courses you can take at your own pace. These courses often include video lessons, quizzes, and interactive exercises.

**Language exchange**

Joining a language exchange group allows you to practice Thai with native speakers while helping them with English. Meetup.com and local Facebook groups are good places to find language exchange events. This method not only helps you practice speaking but also provides cultural insights and social connections.

**Self-study**

For those who prefer self-guided learning, there are plenty of resources available. Books like 'Teach Yourself Thai' and 'Thai for Beginners' provide structured lessons and exercises. Additionally, YouTube channels like **Learn Thai with Mod** and **ThaiPod101** offer free video tutorials on various aspects of the language.

### Immersive experiences

Immersing yourself in the local culture is one of the most effective ways to learn a language. Engage in daily conversations with locals, visit markets, attend cultural events, and participate in community activities. The more you immerse yourself, the faster you'll learn.

### On the street

Practical, real-world practice is invaluable. Use your Thai skills while shopping, dining out, or travelling. Simple interactions, like ordering food or asking for directions, provide practical experience and help reinforce your learning.

### Language meetups and clubs

Many cities in Thailand have language meetups and clubs where expats and locals gather to practice different languages. Joining these groups can provide a supportive and social environment for practising Thai.

### Cultural centres

Institutions like the Siam Society and various cultural centres often offer language classes and cultural workshops. These centres provide a holistic learning experience, combining language instruction with cultural education.

## Study tips for learning Thai

Here are some practical tips to help you effectively learn and master the Thai language.

### Start with the basics

Begin by learning essential phrases and greetings. Mastering basics like 'hello' (*sawasdee*), 'thank you' (*khob khun*), and 'excuse me' (*khor thot*) can go a long way. Focus on common words and phrases used in everyday conversations, as these will be the most practical and useful in your daily interactions.

**Practice regularly**

Consistency is key when learning a new language. Set aside a specific time each day for language practice. Even 15-20 minutes a day can significantly impact over time. Regular practice helps reinforce what you've learned and makes it easier to build upon your existing knowledge.

**Engage in active listening**

Listening to native speakers helps you get accustomed to the sounds and rhythm of the language. Watch Thai TV shows and movies, and listen to Thai music or podcasts. Try to mimic the pronunciation and intonation you hear, as this will improve your speaking skills and help you sound more natural.

**Use language apps**

There are numerous language learning apps like **Duolingo**, **Mondly**, and **Pimsleur** that offer structured lessons in Thai. These apps provide interactive exercises, quizzes, and pronunciation guides to help you learn effectively. They also allow you to practice on the go, making it easier to fit language learning into your daily routine.

**Label your surroundings**

A fun and practical way to learn vocabulary is by labelling items around your home with their Thai names. This constant exposure helps reinforce your memory. Each time you see the label, it serves as a mini-review session, helping you to internalise the vocabulary.

**Practice speaking**

Don’t be afraid to speak, even if you make mistakes. Practice speaking Thai with locals, friends, or language exchange partners. The more you practice, the more confident you’ll become. Speaking regularly helps you become more comfortable with the language and improves your fluency over time.

**Learn the Thai script**

While it might initially seem daunting, learning the Thai script can be incredibly rewarding. It helps with pronunciation and understanding the structure of the language. Start with basic letters and gradually move to more complex characters. Learning the script also allows you to read signs, menus, and other written materials, immersing you in the language.

**Join language exchange groups**

Language exchange groups provide an opportunity to practice Thai with native speakers while helping them with English. It's a win-win situation and a great way to make friends. Engaging in these exchanges can also offer cultural insights and make the learning process more enjoyable and social.

**Use visual aids**

Flashcards, pictures, and charts can be helpful tools for memorising vocabulary and phrases. Visual aids make learning more engaging and are particularly useful for visual learners. They can break the monotony of traditional study methods and make the learning process more dynamic.

**Stay patient and positive**

Language learning is a gradual process, and it's natural to encounter challenges along the way. Stay patient and keep a positive attitude. Celebrate small milestones and progress. Remember that every bit of progress, no matter how small, is a step towards fluency. Maintaining a positive outlook will keep you motivated and persistent.

**100 everyday phrases and words**

Mastering everyday phrases and slang can significantly enhance your communication ability in Thailand. Here are 100 essential words and phrases to help you 'survive' and thrive:

| **Greetings** | **Politeness** |
|---|---|
| 1. Hello: *Sawasdee*<br>2. Good morning: *Aroon sawat*<br>3. Good evening: *Sawatdee ton chao*<br>4. Good night: *Ra tree sawat*<br>5. Goodbye: *Laa gorn*<br>6. How are you?: *Sabai dee mai?*<br>7. I'm fine, thank you: *Sabai dee, khob khun*<br>8. Nice to meet you: *Yin dee tee dai ruu jak*<br>9. What's your name?: *Khun chue arai?*<br>10. My name is…: *Chan chue...* | 11. Thank you: *Khob khun*<br>12. Excuse me/sorry: *Khor thot*<br>13. Please: *Karuna*<br>14. Yes: *Chai*<br>15. No: *Mai chai*<br>16. You're welcome: *Mai pen rai*<br>17. No problem: *Mai mee pan ha*<br>18. Good luck: *Chok dee*<br>19. Happy birthday: *Suk san wan gert*<br>20. Congratulations: *Yin dee duay* |

| **Common phrases** | **Basic questions** |
|---|---|
| 21. I don't understand: *Chan mai kao jai*<br>22. Do you speak English?: *Pood pasa ang-grit dai mai?* | 31. How much?: T*ao rai?*<br>32. Where is…?: *Tee nai?*<br>33. What is this?: *Nee arai?*<br>34. When?: *Meu rai?* |

| | |
|---|---|
| 23. I need help: *Chan tong gaan khwam chuai*<br>24. I'm lost: *Chan long tang*<br>25. Call the police!: *Riak dtam-ruat!*<br>26. Can you help me?: *Chuai chan noi dai mai?*<br>27. I'm sorry, I don't understand: *Khor thot, chan mai kao jai*<br>28. Please speak slowly: *Karuna pood cha cha*<br>29. I'm learning Thai: *Chan kamlang rian pasa Thai*<br>30. Just a moment: *Ru sak kru* | 35. Why?: *Tum mai?*<br>36. Who?: *Krai?*<br>37. Which?: Anai?<br>38. How?: *Yang ngai?*<br>39. Can I?: D*ai mai?*<br>40. Do you have…?: *Mee… mai?* |

| Travel | Food and drink |
|---|---|
| 41. Taxi: *Taxi*<br>42. Bus: *Rot may*<br>43. Train: *Rot fai*<br>44. Airport: *Sanam bin*<br>45. Hotel: *Rong raem*<br>46. Where is the restroom?: *Hong nam yoo tee nai?*<br>47. Left: *Sai*<br>48. Right: *Kwaa*<br>49. Straight ahead: *Trong* | 51. Water: *Nam*<br>52. Food: *Ahan*<br>53. Menu: *Menu*<br>54. Bill: *Check bin*<br>55. Delicious: *Aroi*<br>56. Spicy: *Phet*<br>57. Not spicy: *Mai phet*<br>58. Vegetarian: *Mangsawirat*<br>59. Can I have the bill?: *Check bin duay ka/krub* |

| | |
|---|---|
| *bpai*<br>50. Turn around: *Glap bpai* | 60. I'm allergic to...: *Chan phae...* |

| **Shopping** | **Health** |
|---|---|
| 61. Cheap: *Tuk*<br>62. Expensive: *Paeng*<br>63. just looking: *Doo chao chao*<br>64. Discount: *Lod ra ka*<br>65. Shop: *Ran khai*<br>66. Market: *Talad*<br>67. Can I try it on?: *Kor long dai mai?*<br>68. Do you have another size?: *Mee size eun mai?*<br>69. Can you give a discount?: *Lod ra ka noi dai mai?*<br>70. It's too expensive: *Paeng mak pai* | 71. I need a pharmacy: *Chan tong gaan ran khai ya*<br>72. Hospital: *Rong payabaan*<br>73. Where's the hospital?: *Rong payabaan yoo tee nai?*<br>74. Pharmacy: *Ran khai ya*<br>75. I don't feel well: *Chan mai sabai*<br>76. I am sick: *Chan mai sabai*<br>77. I need a doctor: *Chan tong gaan mor*<br>78. I have a fever: *Chan mee khai*<br>79. Medicine: *Ya*<br>80. Doctor: *Mor* |

| **Slang** | **Numbers** |
|---|---|
| 81. Cool: *Jeen*<br>82. Crazy: *Baa*<br>83. Friend: *Puak*<br>84. Awesome: *Laa* | 91. One: *Neung*<br>92. Two: *Song*<br>93. Three: *Saam*<br>94. Four: *See* |

| | |
|---|---|
| 85. Hurry up: *Reo reo* | 95. Five: *Haa* |
| 86. What's up?: *Ben yang ngai bang?* | 96. Ten: *Sip* |
| 87. No worries: *Mai pen rai* | 97. Twenty: *Yee sip* |
| 88. Take it easy: *Bao bao* | 98. Thirty: *Saam sip* |
| 89. Let's go: *Pai gan* | 99. Hundred: *Roi* |
| 90. Really?: *Jing jing?* | 100. Thousand: *Pun* |

**Conclusion**

Learning Thai as a digital nomad offers many benefits, from enhancing daily interactions and building relationships to gaining cultural appreciation and professional advantages. By starting with the basics, practising regularly, and utilising various study methods, you can make significant progress in your language learning journey. Whether you enrol in a language school, hire a private tutor, or immerse yourself in everyday conversations, the effort you put into learning Thai will be greatly rewarded. Embrace the challenge, stay patient, and enjoy discovering a new language and culture. Happy learning!

**Stay tuned . . .**

In the next chapter, **Travel & Exploration**, we will embark on an exciting journey across Thailand, exploring everything from majestic temples to pristine tropical islands. Whether you're drawn to bustling cities, serene beaches, cultural marvels, or natural wonders, this chapter will guide you through Thailand's must-visit destinations and activities. Discover UNESCO sites, engage in thrilling adventures like trekking, scuba diving, and rock climbing, and find the perfect weekend getaways from major cities like Bangkok, Chiang Mai, Phuket, and Pattaya.

But that's not all! We'll also uncover hidden gems and less crowded spots for those seeking a more off-the-beaten-path experience. Additionally, take advantage of easy long weekend trips to neighbouring countries such as Cambodia, Hong Kong, and Vietnam. To ensure a smooth journey, we'll provide top travel hacks and tips for navigating local transportation, ridesharing apps, and air travel. Stay tuned as we delve into the wonders that await in **Chapter 7**!

# Chapter 7: Travel & Exploration

## Exploring Thailand: From Temples to Tropical Islands

Thailand is a treasure trove of adventures waiting to be uncovered. Whether you're a city slicker, a beach bum, or a culture enthusiast, there's something here for you. From the bustling streets of Bangkok and the serene temples of Chiang Mai to the sun-soaked islands of Phuket and Koh Samui, the Land of Smiles offers endless opportunities for exploration. In this chapter, we'll guide you through must-visit destinations, weekend getaways, and even some hidden gems off the beaten path. Pack your bags and get ready to explore!

## TOP DESTINATIONS & ACTIVITIES

Thailand is a land brimming with vibrant cities, stunning beaches, and rich cultural heritage. Whether you're chasing the pulsing energy of urban life, seeking the peaceful embrace of tropical islands, or yearning to immerse yourself in the deep-rooted traditions of its cultural sites, Thailand has it all. Here, we'll take you on a journey through some of the must-visit places in this beautiful country, ensuring your travel itinerary is packed with unforgettable experiences.

### Must-Visit Places in Thailand

Beaches and Islands

#### Koh Samui

Koh Samui is a tropical paradise known for its palm-fringed beaches, coconut groves, and luxury resorts. **Chaweng Beach** is a hotspot for nightlife, while **Lamai Beach** offers a more relaxed atmosphere. Don't forget to visit the **Big Buddha** and take a day trip to the **Ang Thong National Marine Park**.

### Koh Phi Phi Leh

The world-famous Maya Bay, featured in the iconic film 'The Beach' starring Leonardo DiCaprio, is located on Koh Phi Phi Leh. 100-metre-high limestone cliffs shelter this stunningly beautiful bay on three sides. Note that it is closed annually every August and September to preserve its natural beauty and delicate ecosystem.

### Phang Nga Bay

Phang Nga Bay is renowned for its emerald-green waters and towering limestone karsts. Take a boat tour to explore the dramatic landscape, with highlights including **James Bond Island** and the floating village of **Koh Panyee**.

### Koh Lanta

For a more laid-back island experience, head to Koh Lanta. Known for its long sandy beaches and relaxed atmosphere, it's a great spot for snorkelling, diving, and exploring the **Mu Ko Lanta National Park**.

### Koh Chang

Located near the Cambodian border, Koh Chang is Thailand's second-largest island and offers a mix of beaches, rainforests, and waterfalls. **White Sand Beach** is the most developed, while **Lonely Beach** is popular with backpackers.

## Bustling cities

### Bangkok

The capital city, Bangkok, is a bustling metropolis that never sleeps. From the grandeur of the **Grand Palace** and **Wat Arun** to the vibrant street life and night markets, Bangkok offers a mix of history, culture, and modernity. Don't miss out on the **floating markets** and a ride on a traditional long-tail boat along the **Chao Phraya River**.

**Chiang Mai**

Nestled in the mountainous northern region, Chiang Mai is known for its serene temples, lush landscapes, and a slower pace of life. Visit the iconic **Wat Phra That Doi Suthep** for stunning city views, explore the bustling **Night Bazaar**, and if you're there in November, experience the magical **Yi Peng Lantern Festival**.

**Phuket**

Thailand's largest island is famed for its vibrant nightlife, luxurious resorts, and beautiful beaches. Patong Beach is the heart of the action, while **Kata** and **Karon beaches** offer a more laid-back vibe. Don't miss the **Big Buddha** and the **Old Town's** Sino-Portuguese architecture.

**Pattaya**

Once a quiet fishing village, Pattaya has transformed into a bustling city known for its beaches, water sports, and nightlife. Visit the **Sanctuary of Truth**, a wooden temple with intricate carvings, or take a day trip to the nearby **Coral Island** (Koh Larn) for some snorkelling and relaxation.

**Ayutthaya**

Just a short drive from Bangkok, Ayutthaya is a UNESCO World Heritage site that offers a glimpse into Thailand's rich history. Explore the ancient ruins and temples, such as **Wat Phra Si Sanphet** and **Wat Mahathat**, where you'll find the iconic Buddha head entwined in tree roots.

Cultural marvels

**Wat Pho and Wat Arun, Bangkok**

Wat Pho is home to the famous **Reclining Buddha** and is one of Bangkok's oldest and largest temples. The temple complex is also considered the birthplace of traditional Thai massage, so be sure to indulge in a relaxing session. Wat Arun, also known as the **Temple of Dawn**, is one of Bangkok's most iconic landmarks. Situated on the west bank of the Chao Phraya River, its towering spires are adorned with colourful porcelain and seashells. Visiting at sunset provides a breathtaking view as the temple lights up and reflects on the river.

**Wat Rong Khun (White Temple), Chiang Rai**

Unlike any other temple in Thailand, Wat Rong Khun, or the White Temple, is a contemporary, unconventional creation by artist Chalermchai Kositpipat. Its all-white exterior adorned with mirrors makes it a surreal and stunning sight. Visitors are often captivated by the intricate details and symbolic elements embedded throughout the temple grounds.

**Phanom Rung Historical Park, Buriram**

Located in northeastern Thailand, Phanom Rung Historical Park features an ancient Khmer temple complex built on an extinct volcano. The temple is dedicated to the Hindu god Shiva and offers a unique glimpse into the region's historical and cultural influences. The park is especially renowned for its sunrise and sunset views, which align perfectly with the temple's doorways during certain times of the year.

**Wat Phra That Lampang Luang, Lampang**

Located in Lampang province, Wat Phra That Lampang Luang is one of the best-preserved Lanna-style temples in northern Thailand. The temple complex dates back to the 13th century and is known for its impressive wooden structures, intricate carvings, and the revered **Emerald Buddha** statue. It's a serene and historically rich site that offers a deep dive into Lanna culture.

### Wat Benchamabophit (Marble Temple), Bangkok

Wat Benchamabophit, also known as the Marble Temple, is a stunning example of modern Thai architecture. Built from Italian marble, the temple features an elegant, symmetrical design and beautiful gardens. The interior houses a magnificent Sukhothai-style Buddha statue, a peaceful retreat in the heart of Bangkok.

## Natural wonders

### Similan Islands National Park

Located in the Andaman Sea, the Similan Islands National Park is a marine wonderland. The park consists of 11 islands, each boasting crystal-clear waters, vibrant coral reefs, and abundant marine life. It's a top destination for diving and snorkelling, offering sites like **Elephant Head Rock** and **Christmas Point**. On land, you can enjoy pristine beaches and lush tropical forests.

### Thung Salaeng Luang National Park

Situated in northern Thailand, Thung Salaeng Luang National Park offers a unique landscape of pine forests, grasslands, and rolling hills. It's a great spot for camping, birdwatching, and exploring the park's numerous waterfalls and caves. The park is also home to diverse wildlife, including deer, wild boar, and various bird species.

### Tarutao National Marine Park

Tarutao National Marine Park, located in the southern part of Thailand, comprises 51 islands in the Andaman Sea. The park is known for its rugged landscapes, pristine beaches, and abundant marine life. Activities include snorkelling, diving, and exploring the island's dense forests. **Koh Tarutao**, the largest island, offers historical sites and hiking trails, adding a touch of adventure to your visit.

### Mu Ko Ang Thong National Marine Park

Mu Ko Ang Thong National Marine Park is an archipelago of 42 islands in the Gulf of Thailand near Koh Samui. The park is known for its stunning limestone cliffs, hidden lagoons, and white sandy beaches. Activities include kayaking, snorkelling, and hiking to the viewpoint of **Koh Wua Ta Lap**, which offers panoramic views of the islands. The park's rich marine life and vibrant coral reefs make it a popular destination for divers and snorkelers.

### Doi Suthep-Pui National Park

Located near Chiang Mai, Doi Suthep-Pui National Park is home to the famous **Wat Phra That Doi Suthep** temple, which offers breathtaking views of the city below. The park features lush forests, waterfalls, and diverse wildlife. Visitors can explore the park's numerous hiking trails, visit the Bhubing Palace, and enjoy the natural beauty of the **Mon Tha Than** and **Huay Kaew** waterfalls. The park is also a great spot for birdwatching, with over 300 bird species recorded.

### UNESCO sites

### Ayutthaya Historical Park

Ayutthaya Historical Park, located just north of Bangkok, is a UNESCO World Heritage site that showcases the ruins of the ancient capital of the Kingdom of Ayutthaya. The park is filled with stunning temples, palaces, and statues, each telling a story of the city's glorious past. Highlights include **Wat Phra Si Sanphet**, **Wat Mahathat**, and the iconic Buddha head entwined in tree roots.

### Sukhothai Historical Park

Sukhothai Historical Park, another UNESCO World Heritage site, offers a fascinating glimpse into Thailand's past. The park contains the ruins of the ancient city of Sukhothai, which was the capital of the first Kingdom of Siam. Wander through the impressive temples, such as **Wat Mahathat** and **Wat Si Chum**, and learn about the history and culture of the Sukhothai Kingdom.

**Ban Chiang Archaeological Site**

Located in northeastern Thailand, the Ban Chiang Archaeological Site is recognised as one of the most important prehistoric settlements in Southeast Asia. The site provides evidence of early agricultural practices, metallurgy, and social organisation. Visitors can explore the **Ban Chiang National Museum**, which houses artefacts and exhibits detailing the site's historical significance.

**Thungyai-Huai Kha Khaeng Wildlife Sanctuaries**

The Thungyai-Huai Kha Khaeng Wildlife Sanctuaries, located in western Thailand, form a UNESCO World Heritage site renowned for its rich biodiversity. These sanctuaries cover a vast area of tropical forests and are home to numerous endangered species, including tigers, elephants, and gibbons. The sanctuaries offer limited access to visitors, but guided tours can provide an opportunity to explore this pristine natural habitat.

**Dong Phayayen-Khao Yai Forest Complex**

The Dong Phayayen-Khao Yai Forest Complex, which includes **Khao Yai National Park**, is a UNESCO World Heritage site known for its rich biodiversity and stunning landscapes. The complex spans over 230 km and is home to a diverse range of flora and fauna, including elephants, tigers, and hornbills. The park features lush forests, grasslands, and waterfalls, with numerous trails and viewpoints offering spectacular opportunities for hiking and wildlife spotting.

## Must-Try Activities in Thailand

Ready to get your adrenaline pumping and immerse yourself in Thailand's stunning natural landscapes? From trekking through lush jungles and rafting down wild rivers to diving into vibrant underwater worlds and cycling along scenic routes, Thailand offers a playground for outdoor enthusiasts. Whether you're a seasoned adventurer or a newbie looking to try something new, there's an exciting activity waiting for you. Let's dive into some of the best outdoor pursuits in this diverse and beautiful country.

### All-terrain vehicle (ATV)

#### Chiang Mai ATV Tours

Chiang Mai's mountainous terrain provides the perfect backdrop for off-road adventures. Various tour operators offer ATV excursions that take you through forests, rivers, and hill tribe villages. It's an exciting way to explore the region's natural beauty and cultural heritage. You can opt for half-day or full-day tours, which often include stops at waterfalls and scenic lookouts. The tours usually provide all necessary equipment and safety briefings, ensuring a fun and safe adventure. Check out **8 Adventures** and **X-Centre Chiang Mai**.

#### Kanchanaburi ATV Tours

Kanchanaburi, famous for the Bridge over the River Kwai, offers exciting off-road adventures. Ride through dense jungles, along riverbanks, and up to mountainous areas. The tours often include visits to historical sites, local villages, and scenic spots. Kanchanaburi's diverse terrain makes it an ideal destination for ATV enthusiasts looking for adventure and cultural experiences. Check out **CRD Adventure** and **River Kwai Village**.

**Pai Off-Road Adventures**

Pai, a charming town in northern Thailand, is known for its stunning landscapes and laid-back vibe. Off-road adventures in Pai take you through lush valleys, bamboo forests, and along the Pai River. The trails vary in difficulty, catering to both beginners and experienced riders. Along the way, you'll encounter hot springs, waterfalls, and breathtaking viewpoints. ATV tours in Pai offer a perfect blend of adventure and natural beauty. Check out **SSJ ATV PAI** and **Bon Voyage**.

**Pattaya Off-Road Safari**

Pattaya offers thrilling ATV and off-road adventures through its diverse landscapes. Ride through jungles, plantations, and rugged trails, enjoying the adrenaline rush and scenic views. Tours typically include safety gear, instructions, and guides to lead the way. You can choose from short trips to full-day excursions, making it suitable for beginners and seasoned adventurers. The off-road trails often lead to hidden gems like secluded beaches and panoramic viewpoints. Check out **Pattaya Off-Road Safari** and **ATV Pattaya Jungle Adventures**.

**Phuket ATV Adventures**

Phuket is not just about beaches; it also offers fantastic off-road adventures. Explore the island's interior on an ATV, navigating through rubber plantations, tropical forests, and rugged hills. Tours often include visits to local farms, traditional villages, and hidden waterfalls. ATV adventures in Phuket provide a unique perspective of the island's less-explored areas, making for an unforgettable experience. Check out **Phuket ATV Tour** and **Phuket Paradise Trip ATV Adventure.**

## Cycling

### Chiang Mai's Old City

Chiang Mai's Old City is an ideal place for a leisurely bike ride. The area is relatively flat and filled with historical temples, local markets, and charming streets. You can rent a bike for around **THB 50-100 per day** from various shops around the city. Cycling through the Old City allows you to explore at your own pace and discover hidden gems.

### Bang Krachao (Bangkok's Green Lung)

Bang Krachao, also known as Bangkok's Green Lung, is a serene escape from the bustling city. The area features a network of cycling trails through lush greenery and along the Chao Phraya River. You can rent a bike for about **THB 100 per hour** or **THB 200-300 for the entire day**. Alternatively, you can bring your own bike and enjoy the trails for free.

### Phuket's Coastal Roads

Phuket offers several scenic coastal roads perfect for cycling. One popular route is from Patong to Karon Beach, which provides stunning ocean views and a relatively challenging ride due to the hilly terrain. While there are no formalised bike lanes, early mornings offer quieter roads and fewer vehicles. If you have your own bike, this can be a cost-free way to enjoy Phuket's natural beauty.

### Phang Nga Bay

Phang Nga Bay is famous for its dramatic limestone karsts and emerald-green waters. While most people explore the bay by boat, the surrounding areas offer excellent opportunities for cycling. You can ride through local villages, rubber plantations, and along quiet coastal roads. Bringing your own bike or renting one locally can keep costs low, and you'll get a unique perspective on this stunning region.

**Sukhothai Historical Park**

Sukhothai Historical Park is a UNESCO World Heritage site with ancient ruins and beautiful landscapes. The park is well-suited for cycling, offering a peaceful and scenic way to explore the historical ruins. Entry to the park is around **THB 100**, and you can rent a bike for about **THB 30-50 per day**. Cycling through the park allows you to cover more ground and see more sites than you would on foot.

**Tours**

If you fancy joining a tour group for a day, week, or even a month, we recommend the following reputable companies:

- **Grasshopper Adventures** offers a variety of cycling tours throughout Thailand, ranging from day trips to multi-day adventures. Their tours are well-organised and cover some of the most scenic and culturally rich areas of the country.
- **SpiceRoads Cycling** provides various cycling tours, from leisurely rides through Bangkok's backstreets to challenging multi-day journeys through the countryside. They cater to all skill levels and ensure a memorable experience.
- **Bike Tour Asia** specialises in custom and group cycling tours around Thailand and Southeast Asia. They offer both guided and self-guided tours, allowing you to choose the level of support you need.
- **Bangkok Bike Tours** focuses on city cycling tours, giving you a unique perspective on Bangkok's bustling streets and hidden gems. Their knowledgeable guides provide fascinating insights into the city's history and culture.
- **Chiang Mai Bicycle** offers a variety of cycling tours in and around Chiang Mai, including cultural tours, nature

rides, and multi-day excursions. Their tours are designed to show you the best of Northern Thailand.

**Kiteboarding, wakeboarding, and cable skiing**

**Hua Hin**

Hua Hin, located just a few hours from Bangkok, is one of the top spots for kiteboarding in Thailand. The consistent winds and shallow waters create perfect conditions for beginners and experienced kiteboarders. The best time for kiteboarding in Hua Hin is between November and April when the winds are strongest. Several schools, such as **Kiteboarding Asia** and **Surfspot**, offer lessons and equipment rentals, making it easy to get started. Whether you want to learn the basics or perfect your tricks, Hua Hin's welcoming kiteboarding community and beautiful beaches provide the ideal setting.

**Koh Samui / Koh Phangan**

The stunning beaches of Samui and Phangan are another very popular spot for kiteboarding. The beach offers favourable wind conditions and beautiful surroundings, making it a prime spot for kiteboarding enthusiasts. The best months for kiteboarding in Koh Samui are from December to March. There are several kiteboarding schools, such as **ION Club** and **Siam Pro Kite**, where you can take lessons and rent gear. These schools cater to all skill levels, from beginner courses to advanced clinics. With its stunning backdrop and excellent wind conditions, Maenam Beach on Koh Samui is a kiteboarder's paradise.

**Pattaya**

Pattaya, known for its vibrant nightlife, also offers great kiteboarding opportunities. The best spot for kiteboarding in Pattaya is Jomtien Beach, where the wind conditions are ideal from November to April. Schools like **KBA Pattaya** and **Club Loong Chat** provide lessons and equipment rentals, ensuring you have everything to hit the water. The beach is spacious and less crowded than other areas, making it perfect for practising your skills and enjoying the thrill of kiteboarding.

**Phuket**

Phuket, famous for its stunning beaches, is also a fantastic destination for kiteboarding. Nai Yang Beach and Chalong Bay are the top spots for this exhilarating sport. The best time to kiteboard in Phuket is from November to April. Schools such as **Kite Zone** and **KBA Phuket** offer lessons ranging from beginner to advanced levels and equipment rental. The picturesque scenery and reliable wind conditions make Phuket a top choice for kiteboarders looking for adventure and beauty.

**Pranburi**

Just south of Hua Hin, Pranburi is an up-and-coming kiteboarding destination. With fewer tourists, it offers a more relaxed and serene environment for kiteboarding. The winds here are consistent, especially between November and March. Schools like **Kite School Pranburi** and **Kite Thailand** operate in Pranburi and offer lessons and gear rentals. The long, uncrowded beaches and stunning coastal views make Pranburi a hidden gem for kiteboarding enthusiasts seeking a quieter spot.

**Thai Wake Park (Bangkok)**

Located just outside Bangkok, Thai Wake Park offers world-class facilities for wakeboarding and cable skiing. The park features multiple lakes with different cable systems catering to all skill levels, from beginners to advanced riders. Equipment rentals and lessons are available, and there's even a pro shop on site. The park also hosts various competitions and events, making it a hub for wakeboarding enthusiasts.

**Phuket Wake Park (Phuket)**

This is another top destination for wakeboarding and cable skiing. The park boasts a state-of-the-art cable system and a range of obstacles to challenge riders of all levels. Equipment rentals and professional coaching are available, ensuring a great experience for everyone. The park's tropical setting adds to the appeal, making it a perfect spot to enjoy water sports in paradise.

**Rafting**

**White water rafting**

For those who crave the rush of navigating through turbulent waters, white water rafting in Thailand is a must. The northern regions, particularly Chiang Mai and Mae Hong Son, are renowned for their excellent rafting spots. The Mae Taeng River offers rapids ranging from Class II to Class IV, providing an exhilarating experience amidst stunning jungle scenery. Tours typically include safety briefings and experienced guides, ensuring fun and safety. Check out **Siam River Adventures** and **Thai Adventure Rafting**.

**Inner tubing**

If you prefer a more laid-back yet equally enjoyable water adventure, inner tubing down Thailand's rivers is a great option. Pai, a charming town in northern Thailand, is famous for its tubing trips along the Pai River. Float leisurely downstream, taking in the beautiful landscapes and stopping at riverside bars for refreshments. It's a perfect way to relax and soak up the natural beauty. Check out **Pai Tubing** and **Khao Sok River Lodge**.

**Bamboo rafting**

For a unique and traditional experience, bamboo rafting offers a serene journey through Thailand's waterways. This gentle activity is particularly popular in the regions surrounding Chiang Mai, Khao Sok and Kanchanaburi. Glide along calm rivers on a raft made from bamboo poles, steered by skilled local guides. It's a fantastic way to immerse yourself in nature and observe the local wildlife. Check out **Eco Tours Chiang Mai** and **Khao Sok Lake**.

## Rock climbing

### Railay Beach (Krabi)

Railay Beach offers a variety of climbing routes suitable for all skill levels. Several climbing schools and guide services operate in the area, providing equipment rental and instruction. Prices for guided climbs start around **THB 1,500 per half-day**. With its stunning limestone cliffs and beautiful beach backdrop, Railay Beach is a must-visit for any climbing enthusiast. Don't miss the famous routes like 'Humanality' and 'The King and I'. Check out **Real Rocks Climbing School, Krabi Rock Climbers** and **King Climbers**.

### Tonsai Beach (Krabi)

Adjacent to Railay Beach, Tonsai Beach is another climbing hotspot. Known for its laid-back vibe and challenging routes, Tonsai attracts climbers looking for a more rugged experience. The area has numerous overhung routes that are perfect for advanced climbers. If you're on a budget, you can camp or stay in one of the affordable bungalows. Many climbing shops offer gear rentals for around **THB 150-200 per day**. Check out **Basecamp Tonsai** and **Rock Shop**.

### Koh Phi Phi

Koh Phi Phi is not just famous for its stunning beaches and nightlife; it also offers excellent climbing opportunities. The limestone cliffs provide a range of routes, from easy beginner climbs to more challenging ascents. Tonsai Tower is a popular climbing spot on the island, offering routes with breathtaking views of the Andaman Sea. You can rent climbing gear locally for about **THB 200-300 per day**, making it a great option for those travelling light. Check **Spider Monkey Climbing**.

**Crazy Horse Buttress (Chiang Mai)**

For those venturing north, the Crazy Horse Buttress near Chiang Mai offers fantastic climbing in a beautiful natural setting. The crag features over 150 routes, catering to all skill levels. The area is well-maintained, and there are several local climbing schools that offer guided tours and equipment rental. Prices for guided climbs start around **THB 1,200 per half-day**. Don't miss the 'Archway' route for a unique climbing experience. Check out **Chiang Mai Rock Climbing Adventures (CMRCA)** and **The Peak Adventure**.

**Khao Chin Lae (Lopburi)**

Khao Chin Lae in Lopburi is a lesser known but excellent climbing destination. The limestone cliffs offer a range of routes, from easy climbs to more technical challenges. The area is also famous for its resident monkeys, adding a bit of wildlife adventure to your climb. You can bring your own gear or rent it locally for about **THB 100-150 per day**. This spot is perfect for those looking to escape the more crowded climbing areas. Check out the **Thai Climbing Development Association (TCDA)** and **57 Hours**.

**Koh Yao Noi (Phuket)**

Koh Yao Noi, located between Phuket and Krabi, is a hidden gem for climbers. The island offers a quieter, more relaxed climbing experience with routes that range from beginner to advanced. The limestone cliffs provide stunning views of Phang Nga Bay. Local climbing shops offer gear rentals for around **THB 200-300 per day**, and guided tours are available for those new to the sport. Don't miss the 'Big Tree' route for an unforgettable climb. Check out **Koh Yao Travel**.

## Scuba diving

### Similan Islands

The Similan Islands are renowned for their stunning underwater landscapes and diverse marine life. The crystal-clear waters offer excellent visibility, making it a top destination for scuba diving and snorkelling. Day trips and liveaboard options are available, with prices starting around **THB 2,500**. Highlights include encounters with manta rays and whale sharks. The islands are part of a national park, so expect pristine conditions and a variety of dive sites, including the famous Elephant Head Rock and Richelieu Rock. Check out **Similan Diving Safaris** and **Sea Bees Diving**.

### Koh Tao

Koh Tao is one of the most popular diving destinations in Thailand, known for its affordable dive courses and beautiful coral reefs. The island offers numerous dive sites suitable for all levels, from beginners to advanced divers. Snorkelling is also fantastic around Shark Bay and Aow Leuk. Dive courses start around **THB 10,000** for Open Water certification. Popular dive sites include Chumphon Pinnacle and Sail Rock, where you can spot barracudas, groupers, and, occasionally, whale sharks. Check out **Big Blue Diving** and **Crystal Dive**.

### Koh Phi Phi

Koh Phi Phi offers excellent diving opportunities, with its colourful coral reefs and diverse marine life. The dive sites around the islands are suitable for all levels, and the water visibility is generally excellent. Prices for dive trips start around **THB 2,000**. Notable dive sites include Bida Nok and Bida Nai, where you can encounter blacktip reef sharks, turtles, and a variety of reef fish. The vibrant nightlife on the island adds to the overall experience. Check out **Phi Phi Scuba Diving Center** and **Blue View Divers**.

### Hin Daeng and Hin Muang

Located in the Andaman Sea, Hin Daeng and Hin Muang are two of Thailand's top diving sites, famous for their deep drop-offs and abundant marine life. These sites are best suited for advanced divers due to the strong currents and depth. Highlights include large schools of fish, manta rays, and the occasional whale shark. Day trips are available from nearby islands like Koh Lanta, with prices starting around **THB 3,500**. The rich biodiversity and stunning underwater scenery make these sites a must-visit. Check out **Scuba Shack** and **Blue Planet Divers**.

### Richelieu Rock

Richelieu Rock, part of the Surin Islands, is often regarded as the best dive site in Thailand. Known for its incredible biodiversity, this pinnacle in the open sea is a magnet for marine life, including large pelagics like whale sharks and manta rays. The site is suitable for intermediate to advanced divers due to its depth and currents. Liveaboard trips are the best way to experience Richelieu Rock, with prices starting around **THB 4,000**. The site offers an unforgettable diving experience with its vibrant coral gardens and diverse species. Check out **The Junk Liveaboards** and **Sea Dragon Dive Center**.

## Trekking

### Doi Inthanon National Park

Doi Inthanon, known as the 'Roof of Thailand', is the highest peak in the country. The national park in Chiang Mai offers hiking trails, from easy walks to challenging treks. The scenery includes waterfalls, dense forests, and stunning viewpoints. The entrance fee is around **THB 300** for foreigners. Don't miss the Ang Ka Nature Trail for a serene walk through a mossy forest or the Kew Mae Pan Trail for breathtaking views. The park is also home to diverse bird species, making it a haven for birdwatchers.

### Khao Sok National Park

Khao Sok National Park, located in southern Thailand, is famous for its dramatic limestone cliffs, dense rainforests, and emerald-green lakes. Guided trekking tours offer a chance to explore the park's diverse flora and fauna. The entrance fee is approximately **THB 200** for foreigners. Keep an eye out for the rare Rafflesia flower and enjoy a boat trip on Cheow Lan Lake for an unforgettable experience. The park also offers excellent opportunities for kayaking and canoeing.

### Khao Yai National Park

Khao Yai National Park is a UNESCO World Heritage site and one of the best places in Thailand for wildlife spotting. Just a few hours north of Bangkok, the park is home to elephants, gibbons, hornbills, and even tigers. The entrance fee is around **THB 400** for foreigners. Guided tours offer a chance to see wildlife in their natural habitat.

### Tours

If you fancy joining a tour group for a day, week, or even a month, we recommend the following reputable companies:

- **Intrepid Travel** offers a variety of trekking tours in Thailand, ranging from day trips to multi-day adventures. Their tours are well-organised and cover some of the most scenic and culturally rich areas of the country.
- **G Adventures** provides a wide range of trekking tours, from leisurely walks to challenging multi-day journeys through the countryside. They cater to all skill levels and ensure a memorable experience.

- **Thailand Hilltribe Holidays** specialises in custom and group trekking tours around Northern Thailand. They offer both guided and self-guided tours, allowing you to choose the level of support you need.
- **Trek Thailand** focuses on trekking tours, giving you a unique perspective on Thailand's natural beauty and hidden gems. Their knowledgeable guides provide fascinating insights into the country's history and culture.
- **Chiang Mai Trekking** offers a variety of trekking tours in and around Chiang Mai, including cultural tours, nature hikes, and multi-day excursions. Their tours are designed to show you the best of Northern Thailand.

**Ziplining**

**Flight of the Gibbon (Chiang Mai)**

One of the most popular ziplining operators, Flight of the Gibbon, offers an unforgettable adventure in the rainforests near Chiang Mai. The course includes over 30 different platforms, sky bridges, and abseils, allowing you to soar through the treetops and catch glimpses of gibbons in their natural habitat. The tour also includes a guided trek and a traditional Thai meal. There is also a sister operation in Phuket.

**Tarzan Adventure (Pattaya)**

For those in southern Thailand, this Pattaya zipline adventure offers an exciting way to explore the region's tropical landscapes. The course features multiple ziplines, sky bridges, and abseils, providing a comprehensive aerial adventure. The experienced guides ensure safety while you focus on enjoying the ride and the stunning views. stunning views along the route too.

**Jungle Flight (Koh Phangan)**

If you find yourself on the island of Koh Phangan, take a break from the beach and experience the Koh Phangan Jungle Flight. This zipline course offers a thrilling journey through the island's lush jungle canopy. The course includes several ziplines, sky bridges, and abseils, offering panoramic views of the surrounding forest and coastline.

**Tree Top Adventure Park (Krabi)**

Tree Top Adventure Park in Krabi offers an exhilarating ziplining experience set amidst lush tropical forests. The park features a variety of zipline routes, sky bridges, and challenging obstacle courses designed to test your agility and bravery. Suitable for both beginners and experienced adventurers, the park ensures safety with professional guides and high-quality equipment. In addition to ziplining, you can enjoy other activities such as rope walking and Tarzan swings. The combination of adventure and stunning natural surroundings makes this a must-visit for thrill-seekers.

**Tarzan Adventure (Phuket)**

For a unique and exhilarating ziplining experience, head to Tarzan Adventure in Phuket. Nestled within the island's lush forest, this zipline park offers a variety of courses that cater to all skill levels. With multiple ziplines, sky bridges, and abseils, you can swing through the treetops like Tarzan himself. The park is designed to provide excitement and safety, with experienced guides and top-notch equipment. This adventure zipline is particularly good for families and children (5+ or 120cm).

**Note:** All zipline adventures are priced from approximately **THB 1,000–3,000** depending on whether it's a half- or full-day package and whether or not transport is required.

### Conclusion

Thailand is truly an adventurer's paradise, offering an incredible range of outdoor pursuits to suit every taste and thrill level. Whether trekking through lush national parks, diving into vibrant underwater worlds, or soaring through the treetops on a zipline, the opportunities for exploration and excitement are endless. The country's diverse landscapes and welcoming atmosphere make it an ideal destination for seasoned adventurers and those looking to try something new. So, pack your gear, embrace the spirit of adventure, and discover the exhilarating outdoor activities that Thailand has to offer.

## WEEKEND GETAWAYS

Thailand is truly an adventurer's paradise, offering an incredible range of outdoor pursuits to suit every taste and thrill level. Whether trekking through lush national parks, diving into vibrant underwater worlds, or soaring through the treetops on a zipline, the opportunities for exploration and excitement are endless. The country's diverse landscapes and welcoming atmosphere make it an ideal destination for seasoned adventurers and those looking to try something new. So, pack your gear, embrace the spirit of adventure, and discover the exhilarating outdoor activities that Thailand has to offer.

### Ideal Short Trips from Major Cities

Discover quick getaways from Thailand’s major cities. Enjoy cultural, natural, and adventurous experiences just a few hours away, perfect for weekend escapes or short breaks from urban life.

#### From Bangkok

#### Kanchanaburi

Just a two-hour drive from Bangkok, Kanchanaburi is a tranquil retreat known for its lush landscapes and historical significance.

Visit the Bridge over the River Kwai, a poignant reminder of World War II, and explore the nearby **Hellfire Pass**. For nature lovers, **Erawan National Park**, with its stunning seven-tiered waterfall, is must-see. Accommodation ranges from riverfront resorts to charming guesthouses, making it a versatile destination for all budgets.

**Hua Hin**

A favoured getaway for Bangkok residents, Hua Hin is a seaside town just three hours away by car. The town offers a mix of beautiful beaches, vibrant night markets, and sumptuous seafood. Visit the **Hua Hin Railway Station**, one of Thailand's oldest, and take a stroll through the **Cicada Market** for handmade crafts and live performances. Don't miss the nearby Phraya Nakhon Cave, home to a stunning royal pavilion bathed in natural light.

**From Chiang Mai**

**Pai**

A three-hour drive through scenic mountain roads brings you to Pai, a laid-back town in northern Thailand that's perfect for a relaxing weekend. Known for its bohemian vibe, Pai offers hot springs, waterfalls, and a vibrant night market. Visit the **Pai Canyon** for breathtaking views, explore the Tham Lod Cave, and unwind in the **Tha Pai Hot Springs**. Pai's eclectic mix of guesthouses, hostels, and boutique resorts ensures something for everyone.

**Chiang Rai**

About three hours from Chiang Mai, Chiang Rai offers a mix of cultural and natural attractions. Visit the iconic **White Temple** (Wat Rong Khun), **Blue Temple** (Wat Rong Suea Ten) and **Black Temple** (Baan Dam Museum) for a dose of contemporary Thai art. The **Golden Triangle**, where Thailand, Laos, and Myanmar meet, is another must-visit. For a unique experience, take a boat ride on the Mekong River. Chiang Rai's serene atmosphere and rich history make it a perfect weekend escape.

### From Phuket

#### Phang Nga Bay

Phang Nga Bay is renowned for its emerald-green waters and dramatic limestone cliffs. Take a day trip from Phuket to explore the bay's numerous islands and caves. Highlights include **James Bond Island** and the floating village of **Koh Panyee**. Kayaking through the bay's hidden lagoons and caves offers a unique perspective of this natural wonder. Various tour operators offer day trips, making it a convenient and memorable weekend getaway.

#### Khao Sok

**Khao Sok National Park** is a nature lover's dream, featuring one of the oldest rainforests in the world, breathtaking limestone karsts, and the stunning **Cheow Lan Lake**. Just a few hours from Phuket, the park offers diverse activities such as jungle trekking, cave exploration, and kayaking. Overnight stays in floating bungalows on the lake provide a unique experience, while guided tours enhance your chances of spotting wildlife like gibbons, elephants, and exotic birds. It's a perfect destination for those seeking adventure and tranquillity in a pristine natural setting.

### From Pattaya

#### Koh Samet

A two-hour drive and a short ferry ride from Pattaya, Koh Samet is a perfect island getaway. Known for its crystal-clear waters and white sandy beaches, the island offers a range of activities, from snorkelling and diving to simply lounging on the beach. The nightlife is vibrant yet laid-back compared to

Pattaya, with beachfront bars and restaurants serving fresh seafood.

### Khao Yai National Park

Just a three-hour drive from Pattaya, Khao Yai National Park is a UNESCO World Heritage site offering lush forests, waterfalls, and diverse wildlife. It's a haven for nature lovers and adventure seekers. Enjoy hiking, wildlife spotting, and visiting the park's beautiful waterfalls, such as **Haew Suwat** and **Haew Narok**. The park also has several vineyards and farms nearby, perfect for a leisurely day of wine tasting and fresh produce sampling.

## Hidden Gems and Less Crowded Spots

While Thailand's major attractions draw the crowds, there are plenty of lesser-known destinations that offer tranquillity and charm. These hidden gems provide a perfect escape from the hustle and bustle, allowing you to experience the country's beauty in a more relaxed and intimate setting.

## From Bangkok

### Koh Si Chang

Just a two-hour drive and a short ferry ride from Bangkok, Koh Si Chang is a serene island that remains relatively untouched by mass tourism. The island offers beautiful beaches, crystal-clear waters, and a relaxed atmosphere. Explore the island's historical sites, such as the **Chao Por Khao Yai Shrine** and the **Summer Palace of King Rama V**. Don't miss the chance to snorkel or dive in the island's pristine waters. Accommodation options range from charming guesthouses to boutique resorts.

### Phetchaburi

A two-hour drive from Bangkok, Phetchaburi is a charming town known for its historical temples and royal palaces. Visit the **Phra Nakhon Khiri Historical Park**, also known as Khao

Wang, which features a palace complex on a hilltop with stunning views. Explore the **Tham Khao Luang Cave**, home to a large reclining Buddha statue and beautiful stalactites. The town's laid-back vibe and rich history make it an ideal weekend getaway.

## From Chiang Mai

### Mae Hong Son

Located five hours north of Chiang Mai, **Mae Hong Son** is a hidden gem known for its stunning landscapes and cultural diversity. Mountains surround the town and offer a range of activities, from trekking to visiting hill tribe villages. Explore the beautiful **Pai River**, the serene **Wat Phra That Doi Kong Mu**, and the picturesque **Pang Ung Lake**. Mae Hong Son's remote location ensures a peaceful and authentic experience.

### Lampang

Just two hours from Chiang Mai, Lampang is a quaint town known for its well-preserved Lanna architecture and traditional horse-drawn carriages. Visit the **Wat Phra That Lampang Luang**, one of the best-preserved Lanna-style temples in northern Thailand. Explore the charming streets of the old town and take a relaxing ride in a horse-drawn carriage. Lampang's laid-back atmosphere and historical charm make it a perfect weekend retreat.

## From Phuket

### Koh Phra Thong

A three-hour journey involving a drive and a boat ride from Phuket, Koh Phra Thong is an undiscovered paradise known for its unspoiled beaches and diverse wildlife. The island offers a unique landscape of savannah-like plains and mangrove forests. It's an ideal destination for birdwatching, snorkelling, and kayaking. The island's eco-friendly resorts and bungalows provide a comfortable stay while preserving the natural beauty.

### Ranong

A three-hour drive from Phuket, Ranong is a hidden treasure known for its hot springs and lush rainforests. Visit the **Raksawarin Hot Springs** and relax in the therapeutic mineral waters. Explore the Ngao Waterfall National Park and the stunning **Laem Son National Park**, which offers beautiful beaches and diverse marine life. Ranong's tranquil environment and natural attractions make it an ideal escape from the crowds.

## From Pattaya

### Koh Larn

A short ferry ride from Pattaya, Koh Larn is a small island that offers a peaceful retreat with its clear waters and beautiful beaches. Unlike Pattaya's bustling beaches, Koh Larn's beaches are more relaxed and less crowded. Enjoy snorkelling, parasailing, and simply lounging on the beach. The island's charming atmosphere and easy access make it a perfect weekend getaway.

### Chanthaburi

A two-and-a-half-hour drive from Pattaya, Chanthaburi is a charming town known for its rich history and beautiful landscapes. Visit the **Cathedral of the Immaculate Conception**, the largest Catholic church in Thailand, and explore the town's gem markets. Don't miss the stunning **Namtok Phlio National Park**, which offers beautiful waterfalls and lush greenery. Chanthaburi's blend of cultural and natural attractions makes it an ideal destination for a relaxing weekend.

## Conclusion

These hidden gems and lesser-known destinations offer a peaceful and unique escape from the usual tourist trails, making them ideal for weekend getaways. Whether you want to explore historical sites, relax on pristine beaches, or immerse yourself in nature, these spots provide a perfect blend of tranquillity and adventure. Similarly, short trips from major cities offer diverse cultural, natural, and adventurous experiences just hours away, perfect for quick escapes from urban life. The variety of options – from historical landmarks to serene retreats – ensures something for every traveller. Next time you need to escape the hustle and bustle of city life, consider these hidden treasures and convenient getaways for a memorable and rejuvenating weekend.

# THAILAND TRAVEL TIPS

Travelling doesn't have to break the bank. With a bit of savvy planning and some insider tips, you can explore Thailand and its neighbouring countries without emptying your wallet. From finding the best deals on accommodations and transportation to discovering affordable yet delicious local cuisine, there are numerous ways to stretch your budget further. Additionally, knowing when to travel and how to take advantage of local discounts can significantly reduce your expenses. Whether you're a soloist, a family, or a group of friends looking to save money, these tips will help you navigate your journey more economically and efficiently.

## Top 10 Travel Hacks

### Plan ahead and book early

One of the easiest ways to save money on travel is to plan ahead and book your flights, accommodation, and activities early. Airlines and hotels often offer significant discounts for early bookings. Use fare comparison sites and set price alerts to snag the best deals. Additionally, consider travelling during the shoulder seasons when prices are lower and crowds are thinner.

**Use public transport**

Public transport in Thailand and its neighbouring countries is not only affordable but also a great way to experience local life. In Thailand, buses, trains, and songthaews (shared taxis) are cost-effective ways to get around. For longer journeys, overnight buses and trains can save you a night's accommodation. In cities like Bangkok, the **BTS Skytrain**, **MRT Metro**, and **Airport Rail Link** are cheap and efficient ways to navigate the city.

**Eat like a local**

Street food is not only delicious but also incredibly budget-friendly. In Thailand, you can enjoy a hearty meal for as little as **THB 50** from street vendors. Look for stalls with long queues, as they're usually a sign of good food. Don't be afraid to try local dishes; they're often more affordable and tastier than Western options. Also, visit local markets for fresh produce and snacks.

**Stay in hostels or guesthouses**

Accommodation can be one of the biggest travelling expenses, but it doesn't have to be. Hostels and guesthouses offer budget-friendly alternatives to hotels. Many hostels in Thailand provide private rooms with en-suite bathrooms at a fraction of the cost of a hotel room. Websites like **Hostelworld** and **Booking.com** are great for finding affordable stays. Additionally, consider staying in a homestay for a more authentic experience.

**House/pet sitting**

Another fantastic way to save on accommodation is by house or pet sitting. Many homeowners look for reliable people to care for their homes and pets while away. Platforms like **TrustedHousesitters, HouseCarers,** and **Official Thailand House and Pet Sitting Facebook Group** connect travellers with these opportunities, allowing you to stay for free in exchange for your services. This not only cuts down your travel costs but also provides a more homely and comfortable stay.

**Travel overnight**

Save on accommodation by travelling overnight. Many buses and trains in Thailand and neighbouring countries offer night services. Not only do you save money on a night's stay, but you also maximise your daytime for exploring. Just make sure to book in advance, especially during peak travel seasons.

**Use local SIM cards**

Avoid expensive roaming charges by purchasing a local SIM card. In Thailand, SIM cards are cheap and offer generous data packages. You can buy them at the airport or local convenience stores. Staying connected allows you to use maps and translation apps and stay in touch with fellow travellers without worrying about hefty phone bills.

**Take advantage of free activities**

Not everything worth doing has to come with a price tag. Many destinations in Thailand offer free or low-cost activities. Explore public parks, visit local temples, and enjoy the beaches. In cities like Bangkok and Chiang Mai, free walking tours are available and provide a great way to learn about the city's history and culture.

**Bargain wisely**

Haggling is a common practice in markets across Southeast Asia. While it might feel uncomfortable at first, it's expected and can save you a lot of money. Start by offering half the asking price and negotiate from there. Always be polite and remember that a friendly smile goes a long way.

**Travel with a buddy**

Travelling with a friend can help you save money in many ways. You can share accommodation costs, split meals, and even negotiate better deals for group activities. Plus, having a travel buddy makes the experience more enjoyable and helps in looking out for one another.

**Do your research**

There are many travel-related sites online with answers to all your questions. You can explore these sites for info about your destination, travel tips, accommodation options, and more. Check out **Culture Trip**, **Lonely Planet**, **Nomads.com**, **Rough Guides**, **Thailand: Off the Beaten Path, Thailand Tourism Directory**, **Travelfish**, **Tripadvisor**, and **Wikitravel**.

## Navigating Transportation Options

Getting around Thailand is an adventure in itself. From tuk-tuks and songthaews to high-speed trains and ferries, the transport options are as diverse as the destinations. Here's a guide to help you navigate the various modes of transport, ensuring you travel smoothly and efficiently.

## Local transport

### Buses

Buses are a cost-effective way to travel both within cities and between destinations. In Thailand, the **Bangkok Mass Transit Authority (BMTA)** operates an extensive network of buses. For intercity travel, there are various classes of buses, from basic to VIP, with reclining seats and air conditioning. Companies like **Green Bus**, **Nakhonchai Air**, and **Bus Online Ticket** offer reliable services. For shorter trips within cities, local buses are an economical option, though they can be crowded and may not have air conditioning.

**Ferries and speedboats**

With its numerous islands, Thailand has an extensive network of ferries. Companies like **Lomprayah** and **Seatran** offer high-speed catamaran services to popular islands like Koh Samui, Koh Phangan, and Koh Tao. For a more leisurely journey, opt for a traditional long-tail boat ride in places like Krabi and Phang Nga Bay. These rides not only provide transport but also a unique way to experience Thailand's beautiful coastal scenery. As the boating industry in Thailand doesn't have the best track record, always go with a reputable company and secure a life jacket once on board. For those traveling to and from Phuket, popular speedboat and ferry companies include **Phuket Ferry**, **Andaman Wave Master**, **Tigerline Ferry**, and **Klook**.

**Minivans**

Minivans are a popular mode of transport for medium to long-distance travel, especially between towns and cities in Thailand. They are faster than buses and offer a more intimate travel experience. Minivans usually depart from major bus terminals and can be booked in advance or on the day of travel. Fares are reasonable (typically **THB 100-300** depending on the distance), and they are a convenient option for reaching destinations not directly accessible by train or large buses. Popular companies include **12Go Asia**, **Bell Travel Service**, and **Thai to Siam**.

**Motorbike taxis**

In busy cities like Bangkok, motorbike taxis are a popular choice for beating traffic. You'll find drivers wearing colourful orange vests at street corners. They're quick and relatively cheap (**THB 10-50**), but always wear a helmet and hold on tight! Remember to negotiate the fare before starting your journey or ensure the driver uses a meter if available.

### Taxis

Taxis are widely available in cities and towns across Thailand. They can be a convenient and comfortable way to get around, especially for longer distances within urban areas. Be sure to check that the driver turns on the meter when you set off, as not doing so is a common scam. If the driver refuses, it's best to get out and find another taxi. Taxi-hailing apps like Grab also offer the convenience of booking a taxi with a set fare.

### Trains

Thailand's train network is a scenic and comfortable way to travel long distances. The **State Railway of Thailand (SRT)** offers various classes, including sleeper trains for overnight journeys. The train from Bangkok to Chiang Mai is particularly popular, offering stunning countryside views. Book tickets in advance, especially during peak travel seasons. For more scenic routes, consider the Eastern & Oriental Express, which offers luxurious journeys through Thailand and neighbouring countries.

### Tuk-tuks and songthaews

Tuk-tuks and songthaews are quintessential modes of transport in Thailand. Tuk-tuks are three-wheeled motorised vehicles that are great for short trips around the city. Always agree on the fare before starting your journey to avoid any surprises. Songthaews are shared pick-up trucks with benches in the back, offering an affordable and fun way to get around. They operate on fixed routes (typically **THB 10-20**), but you can hire them for private trips. In some cities like Chiang Mai, songthaews operate like shared taxis, picking up and dropping off passengers along the route.

## Ridesharing apps

### Grab

Grab is Southeast Asia's leading ride-hailing app, and it's widely used in Thailand. It's a convenient and often cheaper alternative to traditional taxis. You can book cars, motorbikes, and even food delivery through the app. Grab also provides fare estimates upfront, so you know what to expect.

### InDrive

InDrive is another popular ride-sharing app in Thailand. What sets InDrive apart is its unique feature that allows passengers to negotiate fares with drivers directly. This can be a great way to save money, especially for longer trips. The app is user-friendly and offers various vehicle options to suit your needs.

### Bolt

Bolt is a relatively new player in the Thai ride-sharing market, but it's quickly gaining popularity. Known for its competitive pricing, Bolt often has lower fares compared to other ride-hailing services. The app is straightforward to use and provides fare estimates, ensuring no surprises when you reach your destination.

## Air travel

### Budget Airlines

Thailand is well-served by budget airlines like **AirAsia, Nok Air**, and **Thai Lion Air**, making it easy to hop between major cities and tourist destinations. Keep an eye out for promotions and book in advance to get the best deals. Be mindful of baggage allowances to avoid extra fees.

### Domestic Flights

For longer distances within Thailand, domestic flights can save you time. Airports like **Suvarnabhumi** and **Don Mueang** in

Bangkok are hubs for both full-service and budget carriers. Frequent flights connect Bangkok to cities like Chiang Mai, Phuket, and Krabi, making air travel a convenient option for exploring the country. Consider checking secondary airports for more flight options and potentially better deals.

**Conclusion**

Navigating Thailand doesn't have to break the bank. With savvy planning and insider tips, you can explore this beautiful country affordably. Opt for weekend getaways from major cities for quick cultural, natural, and adventurous escapes. Discover hidden gems that offer relaxed, crowd-free experiences. Whether zipping through city traffic on a motorbike taxi, enjoying scenic train rides, or island-hopping on ferries, the journey is just as exciting as the destination. Embrace the adventure, save money, and uncover Thailand's hidden treasures for a memorable trip. Happy exploring!

## EXPLORING NEIGHBOURING COUNTRIES

Thailand is not just a fantastic destination but also a perfect gateway to explore its equally captivating neighbours. With some planning, you can easily hop across borders and experience the rich diversity of Southeast Asia. Whether it's the ancient temples of Cambodia, the serene landscapes of Laos, the bustling cities of Vietnam, the vibrant cultures of Malaysia, or the historical richness of Myanmar, each country offers its own unique allure. These neighbouring destinations are just a short flight or a scenic drive away, making them convenient options for an extended holiday or a spontaneous adventure. Here are some top picks for easy trips to Cambodia, Hong Kong, Indonesia, Laos, Malaysia, Myanmar, Singapore and Vietnam.

## Easy Long-Weekend Trips

## Cambodia

### Siem Reap

Just an hour's flight from Bangkok, Siem Reap is the gateway to the awe-inspiring **Angkor Wat** temple complex. This UNESCO World Heritage site is one of the most significant archaeological wonders in the world. Spend a few days exploring the temples, cycling around the ancient ruins, and enjoying the vibrant nightlife in Siem Reap town. The **Angkor Night Market** is a must-visit for souvenirs and street food.

### Phnom Penh

Cambodia's capital, Phnom Penh, is a fascinating mix of history and modernity. A quick flight from Bangkok, it offers a poignant insight into Cambodia's past with sites like the **Tuol Sleng Genocide Museum** and the **Killing Fields**. On a lighter note, stroll along the riverfront, visit the **Royal Palace** and enjoy the city's burgeoning café scene. Phnom Penh is a city that will leave a lasting impression.

## Hong Kong

### The Island

A short flight from Bangkok, Hong Kong Island is a bustling metropolis known for its impressive skyline and vibrant culture. Explore the **Central district** with its skyscrapers and luxury shops, take a tram ride to **Victoria Peak** for panoramic views, and enjoy the nightlife in **Lan Kwai Fong**. The island offers a mix of modernity and tradition, with temples and street markets coexisting alongside contemporary attractions.

### Kowloon

Kowloon, just across the harbour from Hong Kong Island, offers a different but equally captivating experience. Wander through the bustling streets of **Mong Kok**, visit the famous **Temple Street Night Market**, and take in the views from the Kowloon waterfront. Don't miss the cultural attractions like the **Hong Kong Museum of History** and the **Chi Lin Nunnery**. Kowloon's vibrant street life and rich heritage make it a fascinating destination.

## Indonesia

### Jakarta

Indonesia's capital, Jakarta, is a dynamic city blending modernity with rich cultural heritage. A short flight from Bangkok, the city offers attractions like the **National Monument**, **Taman Mini Indonesia Indah**, and the vibrant markets of **Glodok** (Chinatown). Jakarta's bustling city life, diverse cuisine, and historical landmarks make it a compelling destination for a long weekend.

### Bali

Known for its stunning beaches, lush landscapes, and vibrant culture, Bali is an ideal destination for relaxation and adventure. A short flight from Bangkok, this Indonesian island offers everything from surfing and diving to exploring temples and rice terraces. Visit **Ubud** for its arts and crafts, relax on the beaches of **Seminyak**, and don't miss the iconic **Tanah Lot** temple. Bali's blend of natural beauty and cultural richness makes it a must-visit.

## Laos

### Vientiane

A short flight or an adventurous bus ride from northern Thailand, Vientiane is the laid-back capital of Laos. The city's relaxed atmosphere is perfect for a weekend getaway. Visit the serene **Wat Si Saket**, the grand **Patuxai Monument**, and the sacred That Luang Stupa. Don't miss the riverside night market for delicious Laotian street food and handicrafts.

### Luang Prabang

Nestled in the mountains of northern Laos, Luang Prabang is a UNESCO World Heritage site known for its well-preserved architecture and natural beauty. A short flight from Bangkok or Chiang Mai, this charming town offers a blend of French colonial buildings, ancient temples, and stunning waterfalls. Rise early to witness the alms-giving ceremony, visit the **Royal Palace Museum**, and take a boat trip to the **Pak Ou Caves**.

## Malaysia

### Kuala Lumpur

Malaysia's capital, Kuala Lumpur, is a dynamic city known for its iconic skyline dominated by the **Petronas Twin Towers**. A short flight from Bangkok or Phuket, KL offers a mix of modernity and tradition. Visit the **Batu Caves**, explore the bustling markets of **Chinatown**, and enjoy the vibrant nightlife in **Bukit Bintang**. The city's diverse culinary scene is a treat for food lovers.

### Penang

Penang, known as the 'Pearl of the Orient', is a cultural melting pot with a rich heritage. A quick flight from Bangkok, this island offers a blend of historic architecture, vibrant street art, and delectable cuisine. Wander through the UNESCO-listed **George Town**, visit the **Kek Lok Si** temple, and relax on the beaches of Batu Ferringhi. Penang's hawker food is legendary, so sample as much as possible.

## Myanmar

### Yangon

A short flight from Bangkok, Yangon is Myanmar's largest city and a melting pot of cultures and religions. The city is home to the stunning **Shwedagon Pagoda**, a must-visit landmark that gleams with gold and jewels. Walk around Kandawgyi Lake, explore the colonial-era buildings downtown, and visit the **Bogyoke Aung San Market** for local crafts and jewellery.

### Bagan

For a more historical and spiritual journey, take a short flight from Bangkok to Bagan. This ancient city has over 2,000 temples and pagodas spread across a vast plain. Rent a bicycle and explore the temples at your own pace or take to the skies in a hot air balloon for a breathtaking sunrise view over the pagodas. Bagan's rich history and serene atmosphere make it a truly magical destination.

## Singapore

### Mainland

A quick flight from Bangkok, Singapore offers a perfect blend of modern attractions and cultural heritage. Explore the futuristic **Gardens by the Bay**, visit the historic **Chinatown** and **Little India**, and shop along **Orchard Road**. The city-state's efficient public transport and myriad attractions make for a convenient and exciting getaway.

### Sentosa Island

Just a short ride from mainland Singapore, Sentosa Island is a hub of entertainment and relaxation. Enjoy the beaches, visit **Universal Studios Singapore**, and explore the **S.E.A. Aquarium**. The island offers a range of activities, from adrenaline-pumping adventures to leisurely strolls, making it a versatile destination for all ages.

## Vietnam

### Hanoi

Vietnam's capital, Hanoi, is a city of contrasts, where ancient temples sit alongside bustling markets and modern skyscrapers. A quick flight from Bangkok, Hanoi offers a rich cultural experience. Explore the **Old Quarter's** narrow streets, visit the **Temple of Literature**, and stroll around **Hoan Kiem Lake**. Don't forget to try the local delicacies like pho and banh mi.

### Ho Chi Minh City

Formerly known as Saigon, Ho Chi Minh City is a vibrant metropolis in southern Vietnam. Just a short flight from Bangkok, the city is a hub of activity with its bustling markets, historic landmarks, and lively nightlife. Visit the **War Remnants Museum**, the **Notre Dame Cathedral Basilica**, and the bustling **Ben Thanh Market**. For a day trip, explore the **Cu Chi Tunnels and the Mekong Delta**.

### Conclusion

Exploring neighbouring countries from Thailand is a breeze, with numerous flights, buses, and even boat options available. Whether you're seeking historical insights, cultural experiences, or natural beauty, Cambodia, Hong Kong, Indonesia, Laos, Malaysia, Myanmar, Singapore, and Vietnam offer a treasure trove of adventures just a short trip away. From the ancient temples of Angkor Wat and the serene landscapes of Luang Prabang to the bustling cities of Hanoi and Kuala Lumpur, each destination provides unique experiences. So grab your passport, plan your itinerary, and get ready to explore the wonders that lie beyond Thailand's borders for a memorable journey.

### Stay tuned . . .

You've now delved into the rich tapestry of Thai culture, from understanding its core values to mastering essential etiquette, celebrating vibrant festivals and wrapping your head around the language. As you continue your journey as a digital nomad in the Land of Smiles, there's another exciting aspect waiting to be explored – Thai cuisine!

In **Chapter 8: Food and Dining**, we'll take you on a gastronomic journey through Thailand's diverse and delectable food culture. You'll discover must-try dishes, the best street food spots and tips for navigating local markets. Whether you're a seasoned cook or new to the kitchen, we'll also guide you on shopping for groceries and taking a Thai cooking class. Plus, we'll cover dietary restrictions, ensuring that everyone can enjoy the flavours of Thailand. Get ready to tantalise your taste buds and uncover the culinary delights that make Thai cuisine world-renowned. Stay tuned for an unforgettable adventure in food and dining!

# Chapter 8:
# Food & Dining

### A Culinary Adventure: Tasting Thailand

Welcome to the tantalising world of Thai cuisine, where every meal celebrates flavours, colours, and aromas. From bustling street food stalls to serene local markets, Thailand offers a culinary adventure like no other. Whether you're a fan of spicy curries, aromatic soups, or sweet treats, there's something to delight every palate. This chapter will guide you through the essentials of enjoying Thai food, whether you're dining out, navigating street food, or trying to cook the dishes yourself. Get ready to savour the best of Thailand!

## EXPLORING THAI CUISINE

Thai food is more than just sustenance; it's a way of life that embodies the spirit of community, tradition, and a love for vibrant flavours. If there's one thing you should know about Thai cuisine, balance is key. Each dish is a harmonious blend of sweet, sour, salty, bitter, and spicy elements, creating a symphony of tastes that is both complex and deeply satisfying.

In Thailand, meals are a communal affair. Sharing dishes is the norm, and it's common for families and friends to gather around a table laden with a variety of plates. This practice not only fosters a sense of community but also allows everyone to sample multiple flavours and textures in one sitting. A typical Thai meal includes a variety of dishes such as a curry, a soup, a stir-fry, a salad, and sometimes a dipping sauce, all served with steamed rice. This variety ensures diners experience a wide range of flavours and textures in one meal.

## Overview of Thai Food Culture

Thai cuisine varies significantly by region, each bringing its own unique characteristics to the table. Let's take a look at the four distinct areas.

### Northern

Influenced by neighbouring Myanmar and Laos, Northern Thai cuisine is milder and often features herbs and vegetables. The climate in this region allows for a variety of fresh produce, which is prominently featured in dishes. *Khao soi*, a coconut curry noodle soup topped with crispy fried noodles and pickled mustard greens, is a staple. Another popular dish is *sai oua,* a spicy sausage made with minced pork, herbs, and spices. The use of herbs like dill, mint and cilantro is common, adding a fresh and aromatic quality to the dishes.

### Northeastern

Known for its bold, fiery flavours, Isaan cuisine often includes grilled meats, sticky rice and salads. This region's cuisine is heavily influenced by Lao cooking, which places a strong emphasis on fresh, local ingredients. *Som tum* (papaya salad) is a quintessential dish, combining shredded green papaya, tomatoes, peanuts, dried shrimp and chillies all pounded together in a mortar and pestle. *Larb*, a spicy minced meat salad made with ground rice, fresh herbs, and lime juice, is another iconic dish that packs a punch. Sticky rice, known as *khao niew,* is a staple in Isaan cuisine and is often used to scoop up the various dishes.

### Central

This region is the culinary heart of Thailand, combining flavours from across the country. The fertile plains of Central Thailand produce a wide variety of vegetables, fruits and herbs, which are integral to the local cuisine. Here, you'll find well-known dishes like *pad thai,* a stir-fried noodle dish with shrimp, tofu, peanuts, and bean sprouts and *tom yum goong* (spicy shrimp soup), a hot and sour soup made with lemongrass, kaffir lime leaves, galangal and chillies. *Gaeng keow wan* (green curry) is another popular dish that is made with green curry paste, coconut milk, chicken or beef, and various vegetables. The use of coconut milk and palm sugar in this region's cuisine adds a rich and slightly sweet flavour to the dishes.

**Southern**

Southern Thai food is characterised by its spiciness and the use of coconut milk. The cuisine here is heavily influenced by Malay and Indian cooking, resulting in dishes that are rich, spicy and aromatic. Massaman curry, a rich, slightly sweet curry with peanuts and potatoes and *khao yam*, a colourful rice salad made with fresh herbs, vegetables and toasted coconut, are must-tries. *Gaeng som*, a sour curry made with tamarind and turmeric, is another popular dish. The abundant coastline of Southern Thailand provides a wealth of seafood that is prominently featured in local cuisine.

**The cultural significance of Thai cuisine**

Food is deeply intertwined with Thai culture and traditions. Festivals and celebrations often revolve around food, and certain dishes are prepared for specific occasions. For example, *khao chae*, a dish made with rice soaked in jasmine-scented water and served with various accompaniments, is traditionally eaten during the Songkran Festival (Thai New Year). This dish is not only refreshing but is also believed to have cooling properties, making it perfect for the hot days of the festival. *Kanom jeen*, fermented rice noodles served with various curries, is a popular dish during religious ceremonies and community gatherings. It is often enjoyed in a communal setting, where diners can customise their noodles with a variety of toppings and sauces, highlighting the importance of sharing and community in Thai culture.

Thai cuisine also places a strong emphasis on the use of fresh, local ingredients, which are often sourced from local markets. These markets are bustling centres of activity and serve as social hubs where people gather not only to buy food but also to connect with their neighbours. The vibrant atmosphere of these markets reflects the lively and communal spirit of Thai society.

In Thai households, meal preparation is often a family affair, with multiple generations coming together to cook and share recipes passed down through the ages. This tradition of communal cooking and dining reinforces family bonds and ensures the preservation of culinary heritage.

In summary, Thai food culture is a rich tapestry of flavours, traditions and communal dining. Whether you're enjoying a meal with family, sampling street food or exploring a local market, Thai cuisine offers a culinary adventure that is both delightful and deeply satisfying. So, take your time to savour the flavours, appreciate the balance and immerse yourself in the vibrant food culture of Thailand.

## Popular Dishes and Where to Find Them

Thailand is a paradise for food lovers, with a plethora of dishes that cater to every taste. The country's cuisine is known for its bold flavours, fresh ingredients, and vibrant colours, making every meal an exciting culinary adventure. From street food stalls to high-end restaurants, Thailand offers a diverse range of dining experiences that reflect its rich cultural heritage and regional diversity. Whether you crave something spicy, sweet, sour, or savoury, Thai cuisine has something to tantalise your taste buds. Here are some of thc most popular Thai dishes and the best places to find them, ensuring you get a true taste of Thailand's culinary delights.

**Pad Thai (stir-fried rice noodles)**

Arguably the most famous Thai dish, *pad thai* is a stir-fried noodle dish that combines rice noodles, shrimp, tofu, peanuts, bean sprouts and eggs, all tossed in a tangy tamarind sauce. Originating from Central Thailand, it's a staple of street food culture and a must-try for anyone visiting the country. For an authentic experience, head to **Thip Samai** in Bangkok. This legendary eatery has been serving up mouth-watering pad thai since 1966. The long queues are a testament to its enduring popularity.

**Tom Yum Goong (hot and sour soup with shrimp)**

This spicy and sour shrimp soup is a burst of flavour in every spoonful. Made with fresh shrimp, lemongrass, kaffir lime leaves, galangal and chilli, it's a dish that embodies the complexity of Central Thai cuisine. **Pe Aor** in Bangkok is renowned for its *tom yum goong*, offering a rich, creamy version that is hard to beat. The aromatic broth is both invigorating and satisfying, making it a favourite among locals and tourists alike.

**Som Tum (papaya salad)**

Originating from the Isaan region in Northeastern Thailand, *som tum* is a spicy green papaya salad that is both refreshing and fiery. It typically includes shredded green papaya, tomatoes, peanuts, dried shrimp and chillies, all pounded together in a mortar and pestle. **Somtum Der** in Bangkok is a great place to enjoy this dish in a variety of styles, from the traditional version to more innovative takes. The balance of sour lime, hot chillies, salty fish sauce and sweetness from palm sugar makes this dish a flavour explosion.

**Kaeng Matsaman (massaman curry)**

Massaman curry is a rich, slightly sweet curry that features tender chunks of meat (usually beef or chicken), potatoes and peanuts, all simmered in a fragrant sauce made with coconut milk and a blend of spices. This dish hails from Southern Thailand and is influenced by Malay and Indian cuisines. To taste this comforting dish, visit **Blue Elephant** in Bangkok, a restaurant that has been preserving Thai culinary heritage for decades. The complex blend of spices, including cardamom, cinnamon and star anise, makes this curry a standout.

**Khao Soi (coconut curry noodle soup)**

A Northern Thai speciality, *khao soi* is a coconut curry noodle soup topped with crispy fried noodles, pickled mustard greens and shallots. It's a dish that perfectly balances creamy and spicy elements. **Khao Soi Khun Yai** in Chiang Mai is a beloved spot for this dish, offering a rich and flavourful version that will leave you craving more. The combination of soft egg noodles in a fragrant curry broth and the crunchy fried noodles on top creates a delightful textural contrast.

**Gaeng Keow Wan (green curry)**

Green curry is a staple in Central Thailand, known for its vibrant green colour and aromatic flavour. Made with green curry paste, coconut milk, chicken or beef and a variety of vegetables, it's a dish that is both comforting and invigorating. **Baan Khanitha** in Bangkok is a fantastic place to sample this classic dish in an elegant setting. The green curry paste, made from fresh green chillies, lemongrass and Thai basil, gives the dish its distinctive colour and flavour.

**Khao Niao Mamuang (mango sticky rice)**

No exploration of Thai cuisine would be complete without dessert, and mango sticky rice is a beloved treat. This simple yet delicious dish features ripe mango slices served with sticky rice and drizzled with coconut milk. **Mae Varee** in Bangkok is a popular spot for this dessert, offering perfectly ripe mangoes and a rich, creamy coconut sauce. The combination of the sweet mango, the slightly salty coconut milk and the chewy, sticky rice makes this dish irresistible.

**Gai Pad Med Mamuang (chicken with cashew nuts)**

A delightful stir-fry of chicken, cashew nuts, onions and dried chillies, *gai pad med mamuang* is a dish that combines sweet, salty and spicy flavours. It's a favourite among both locals and visitors. For a great version, try **Sala Rattanakosin Eatery and Bar** in Bangkok, a riverside restaurant known for its stunning views and delicious Thai cuisine. The dish's perfect balance of flavours and the crunchy texture of cashew nuts make it a must-try.

**Laab (minced meat salad)**

*Laab* is a spicy minced meat salad that is a staple in the Isaan region. Made with minced pork or chicken, fresh herbs, lime juice and ground rice, it's a dish that is both flavourful and refreshing. **Laab Ubon** in Ubon Ratchathani is a renowned spot for this dish, offering a variety of versions that highlight the best of Isaan cuisine. The combination of fresh herbs like mint and cilantro, the tangy lime juice and the spicy chillies creates a vibrant and refreshing salad.

**Hoi Tod (oyster omelette)**

A popular street food from Central Thailand, *hoi tod* is a crispy oyster omelette that is both crunchy and savoury. Made with a batter of rice flour and eggs and loaded with fresh oysters, it's often served with a tangy chilli sauce. **Nai Mong Hoi Tod** in Bangkok is famous for this dish, offering a perfect balance of crispy edges and soft, juicy oysters. The bustling atmosphere and the expertly cooked omelettes make it a must-visit for street food enthusiasts.

### Moo Kata (skillet pork)

A unique dining experience originating from Northern Thailand, *moo kata* is a Thai-style barbecue where diners cook their own meat and vegetables on a dome-shaped grill that sits over a pot of soup. It's a communal and interactive way to enjoy a meal. Best enjoyed with friends at local spots like **Best Beef** in Bangkok, where you can grill to your heart's content. The lively ambience and the variety of options make it a perfect spot for a fun and social dining experience.

### Yam Neua Yang (spicy steak salad)

A spicy beef salad from Central Thailand, *yam neua yang* combines grilled beef with fresh herbs, lime juice, fish sauce and chillies. The result is a refreshing and tangy salad with a hint of spice. For an authentic taste, visit **Supanniga Eating Room** in Bangkok, where the beef is perfectly grilled and the salad bursts with flavour. The elegant setting and attentive service add to the overall dining experience, making it a memorable meal.

## Michelin Star Dining in Thailand

Exploring Thai cuisine goes beyond street food and local markets. Believe it or not, Thailand is also home to an impressive array of Michelin-starred restaurants and eateries, offering world-class dining experiences at a fraction of the cost you'd find in Western countries. Here's everything you need to know about Michelin dining in Thailand.

### What is Michelin and why visit?

Michelin is a prestigious culinary guide that awards stars to restaurants based on the quality of their food, service and overall dining experience. Since its inception in France in the early 20th century, the Michelin Guide has been a benchmark for fine dining. Earning a Michelin star is a mark of excellence that can significantly elevate a restaurant's reputation.

**Star ratings**

- **Bib Gourmand:** Recognises eateries offering exceptionally good food at moderate prices, a category perfect for budget-conscious foodies.
- **One Star:** A very good restaurant in its category.
- **Two Stars:** Excellent cooking, worth a detour.
- **Three Stars:** Exceptional cuisine, worth a special journey. As of now, none in Thailand.

**Dining advantages**

- **Affordability:** Michelin-starred dining in Thailand is significantly more affordable than in Western countries. You can experience world-class cuisine without breaking the bank.
- **Cultural Experience:** These restaurants often highlight the best Thai ingredients and culinary traditions, offering a deeper understanding of Thai cuisine. Dishes are crafted with local produce and traditional techniques, providing an authentic and enriching experience.
- **Quality and Innovation:** Michelin-starred chefs in Thailand are known for their creativity and attention to detail, providing a dining experience that is both memorable and unique. They often blend traditional Thai flavours with modern culinary techniques, producing innovative and exquisite dishes.
- **Ambiance:** Many of these restaurants are located in stunning settings, from luxurious hotels to charming colonial houses, adding to the overall dining experience. The ambience is carefully curated to enhance your meal, making it a feast for both the palate and the eyes.

## Where to find Michelin dining in Thailand

Thailand debuted in the Michelin Guide in 2018, and since then, numerous restaurants across the country have earned Michelin stars. As of 2024, Thailand boasts 35 Michelin-starred establishments, spanning Bangkok, Chiang Mai, Phuket, and Phang Nga.

**Bangkok**

- **80/20**
- One Michelin Star. Modern Thai cuisine focussing on seasonal and local ingredients in a modern setting and is located on trendy Charoenkrung Road on the fringes of Chinatown.
- **Price range: THB 2,000-4,000 per person**.
- **Aksorn**
- One Michelin Star. A Thai restaurant and funky rooftop bar offering heritage recipes and traditional culinary techniques in a beautiful setting.
- **Price range: THB 2,000-3,500 per person**.
- **Baan Tepa**
- One Michelin Star. This restaurant has an open-kitchen dining room, chef's table and organic culinary garden that guests can tour before dining to learn about herbs and spices.
- **Price range: THB 3,000-5,000 per person**.
- **Blue by Alain Ducasse**
- One Michelin Star. French cuisine with an emphasis on modern techniques, located in awe-inspiring ICONSIAM on the river.
- **Price range: THB 5,000-8,000 per person**.
- **Cadence by Dan Bark**
- One Michelin Star. Modern cuisine with innovative dishes that showcase global flavours and local ingredients via progressive personal menus.

- **Price range: THB 4,000-6,000 per person**.
- **Canvas**
- One Michelin Star. Offers a unique dining experience through its innovative use of locally sourced ingredients in contemporary cuisine.
- **Price range: THB 3,000-4,500 per person**.
- **Chef's Table**
- Two Michelin Stars. Located in the Lebua complex, it offers a gourmet French dining experience with a panoramic view of Bangkok.
- **Price range: THB 5,000-8,000 per person**.
- **Chim by Siam Wisdom**
- One Michelin Star. This mainstay of the Michelin scene serves up delicious Thai cuisine that combines traditional recipes with modern presentation.
- **Price range: THB 2,000-3,500 per person**.
- **Côte by Mauro Colagreco**
- One Michelin Star. Embracing the Mediterranean, the Riviera-inspired French cuisine of Côte features seasonal ingredients and avant-garde techniques.
- **Price range: THB 4,000-6,000 per person**.
- **Elements, Inspired by Ciel Bleu**
- One Michelin Star. Located in the Okura Prestige Hotel, it offers a unique blend of Japanese and French cuisine in a five-star setting.
- **Price range: THB 4,000-6,000 per person**.
- **Gaa**

  One Michelin Star. This progressive Indian cuisine is well-respected for the way it innovatively and creatively incorporates local Thai ingredients into its dishes.

- **Price range: THB 3,000-5,000 per person**.
- **Haoma**
- One Michelin Star. Indian cuisine with a focus on sustainability and ingredients grown in its own urban farm. Arrive early and take a tour of the gardens before you dine.
- **Price range: THB 2,500-4,000 per person**.
- **IGNIV**
- One Michelin Star. Contemporary cuisine offering a tasting menu-sharing experience with international influences and much to get excited about.
- **Price range: THB 3,000-6,000 per person**.
- **Inddee**
- One Michelin Star. More than just Indian cuisine. Inddee blends traditional flavours with innovative techniques.
- **Price range: THB 3,000-5,000 per person**.
- **Jay Fai**
- One Michelin Star. A famous street food stall offering dishes like crab omelette and drunken noodles. Jay Fai herself is a local legend, and her food is worth the wait.
- **Price range: THB 400-1,000 per dish**. (Note: No reservations, only walk-ins.)
- **Khao (Vadhana)**
- One Michelin Star. Inspired by traditional rice barns, this restaurant serves authentic Thai cuisine with an emphasis on Northern flavours.
- **Price range: THB 2,000-3,500 per person**.
- **Le Du**
- One Michelin Star. Blends Thai flavours with modern culinary techniques. Chef Thitid Tassanakajohn's seasonal menus are a journey through Thailand's diverse cuisine.

- **Price range: THB 2,000-4,000 per person**.
- **Le Normandie**
- Two Michelin Stars. Located in the Mandarin Oriental Hotel, this restaurant offers exquisite French cuisine with stunning views of the Chao Phraya River.
- **Price range: THB 4,000-6,000 per person**.
- **Maison Dunand**
- One Michelin Star. Chef Arnaud Dunand is beloved for his French cuisine, which emphasises contemporary techniques and seasonal ingredients.
- **Price range: THB 3,000-5,000 per person**.
- **Mezzaluna**
- Two Michelin Stars. This restaurant is renowned for its contemporary French cuisine, emphasising seasonal and premium ingredients.
- **Price range: THB 6,000-9,000 per person**.
- **Mia**
- One Michelin Star. Mia offers diners a unique experience: contemporary cuisine focusing on modern European dishes with an Asian twist.
- **Price range: THB 3,000-5,000 per person**.
- **Nahm**
- One Michelin Star. Chef Pim Techamuanvivi showcases refined Thai cuisine with bold flavours, emphasising traditional dishes and regional specialities.
- **Price range: THB 2,500-4,500 per person**.
- **Nawa**
- One Michelin Star. Like many of its contemporaries, Nawa serves Thai cuisine, showcasing local and seasonal ingredients in innovative and unique ways.

- **Price range: THB 2,000-3,500 per person**.
- **Potong**
- One Michelin Star. Potong is the master of Pan-Asian cuisine, serving up creative dishes that blend multiple Asian culinary traditions.
- **Price range: THB 4,500-6,000 per person**.
- **R-Haan**
- Two Michelin Stars. This renowned restaurant emphasises traditional Thai cuisine with contemporary interpretations and surprise twists.
- **Price range: THB 4,000-6,000 per person**.
- **Resonance**
- One Michelin Star. Another spot offering contemporary cuisine that emphasises seasonal and local ingredients. Resonance is well-known for its creative dish presentation.
- **Price range: THB 3,500-5,000 per person**.
- **Samrub Samrub Thai**
- One Michelin Star. Traditional Thai cuisine featuring home-style recipes and hearty flavours. Not as homely as it appears, this place offers sophisticated cuisine.
- **Price range: THB 2,500-4,000 per person**.
- **Saneh Jaan**
- One Michelin Star. Thai cuisine with an emphasis on authentic flavours and regional specialities. Sure to set the taste buds on fire!
- **Price range: THB 2,500-4,000 per person**.
- **Sorn**
- Two Michelin Stars. Specialising in Southern Thai cuisine, Chef Supaksorn 'Ice' Jongsiri uses traditional techniques and local ingredients to create a memorable dining experience.

- **Price range: THB 3,000-5,000 per person**.
- **Signature**
- One Michelin Star. French cuisine emphasising Southern Thai ingredients blended with French techniques in a stunning art-deco dining room.
- **Price range: THB 4,000-6,000 per person**.
- **Sühring**
- Two Michelin Stars. The incredible twin chefs Mathias and Thomas Sühring offer modern German cuisine with various tasting menus that will surprise and delight.
- **Price range: THB 4,000-6,000 per person**.
- **Sushi Masato**
- One Michelin Star. High-end sushi with fish flown in daily from Tokyo's famous Toyosu Market, offering a traditional omakase experience.
- **Price range: THB 4,500-7,000 per person**.
- **Wana Yook**
- One Michelin Star. Thai cuisine with an avant-garde twist, focusing on seasonal and local ingredients, set amidst a charming 100-year-old colonial house.
- **Price range: THB 3,000-5,000 per person.**

**Nonthaburi**

- Suan Thip
- One Michelin Star. Authentic Thai cuisine is served in a beautiful riverside setting. The menu showcases the rich flavours and culinary heritage of Thailand.
- Price range: THB 1,500-3,000 per person.

**Phuket**

- PRU
- One Michelin Star. Located in the Trisara resort, it is known for its farm-to-table approach, using locally

sourced ingredients. Chef Jim Ophorst creates dishes that are both innovative and rooted in Thai tradition.

- **Price range: THB 3,000-5,000 per person**.

**Top Bib Gourmand restaurants and eateries**

Thailand boasts an impressive 196 eateries with Bib Gourmand status, a testament to the country's rich culinary heritage and the diversity of its cuisine. While we can't list them all, here's a selection of some of the best eateries to try across three major cities: Bangkok, Chiang Mai, and Phuket.

**Bangkok**

- **Jay Oh:** A favourite for late-night dining, Jay Oh specialises in seafood dishes like spicy *tom yum soup* and a variety of seafood platters.

**Price range: THB 100-400 per dish.**

- **Likhit Kai Yang:** Specialising in Northeastern Thai barbecue, this eatery is famous for its grilled chicken (*gai yang*) and spicy papaya salad (*som tam*).

**Price range: THB 100-300 per dish.**

- **Prachak Roasted Duck:** Established in 1909, this eatery is renowned for its succulent roasted duck (ped yang). The dish, characterised by its crispy skin and tender meat, is usually served with steamed rice or noodles.

**Price range: THB 100-300 per dish.**

- **Rung Rueang Pork Noodle:** Celebrated for its flavourful pork noodle soups, this restaurant is a go-to for locals and tourists. Options like the pork ball noodle soup and the stewed pork leg soup are highly recommended.

**Price range: THB 60-150 per dish.**

- **Soei:** Known for its bold flavours and authentic recipes, Soei offers a variety of Thai seafood dishes. Highlights include crispy fish with mango salad and spicy crab roe dip.

**Price range: THB 200-500 per dish.**

**Chiang Mai**

- **Huen Phen:** Known for its traditional Northern Thai cuisine, this rustic eatery offers must-try dishes like herbal sausage (*sai oua*) and spicy minced pork salad (*larb*).
- **Price range: THB 100-300 per dish**.
- **Khao Soi Khun Yai:** Famous for its khao soi, a Northern Thai curry noodle soup with a rich and creamy coconut curry base.
- **Price range: THB 50-100 per dish**.
- **Suan Pak:** Offers a farm-to-table experience with dishes made from fresh, locally sourced ingredients. Signature dishes include Northern Thai-style grilled pork and various herbal salads. **Price range: THB 150-300 per dish**.
- **Kiti Panit:** This historic restaurant serves unique takes on traditional Northern Thai dishes, such as *gaeng om* (herbal curry) and *sai ua* (herbal sausage).
- **Price range: THB 120-300 per dish**.
- **Tong Tem Toh:** This popular spot specialises in Northern Thai food, particularly *gaeng hang lay* (Northern Thai curry) and *sai ua* (spicy sausage).
- **Price range: THB 100-300 per dish**.

**Phuket**

**Go Benz Rice Porridge:** Well-known for its late-night *congee* (rice porridge) served with various toppings like century egg and minced pork.

- **Price range: THB 50-150 per dish.**
- **Krua Tungtong:** Celebrated for its fresh seafood and Southern Thai specialties. Dishes like stir-fried crab with curry powder and grilled prawns with garlic are popular choices.
- **Price range: THB 150-400 per dish.**
- **Lock Tien:** A food court-style venue offering a variety of local dishes, including stir-fried noodles with seafood and pork (*hokkien mee*).
- **Price range: THB 50-150 per dish.**
- **Mor Mu Dong:** A beloved local spot known for its authentic Southern Thai dishes, such as spicy shrimp paste dip with vegetables and various curries.
- **Price range: THB 100-300 per dish.**
- **Raya:** This charming restaurant, housed in a colonial building, serves authentic Southern Thai cuisine. The fatty pork is an absolute must-try.
- **Price range: THB 300-800 per dish.**

## Conclusion

Whether you're indulging in street food, dining at a local eatery or experiencing Michelin-starred dining, Thailand's culinary landscape is rich and diverse. Each dish tells a story, reflecting the country's history, culture and the creativity of its people. From the bustling streets of Bangkok to the serene countryside of Chiang Mai, every bite offers a new discovery and a deeper understanding of Thai life and traditions. The array of flavours, the freshness of ingredients and the passion of the cooks make Thai cuisine a truly unforgettable experience. So, grab your chopsticks, spoon and fork and embark on a culinary adventure that will leave you both satisfied and longing for more.

## LOCAL CULINARY EXPERIENCES

Thailand's street food culture is world-renowned, and for good reason. The country's bustling markets and roadside stalls offer a dizzying array of delectable dishes, each more tempting than the last. Street food is not just about grabbing a quick bite; it's an integral part of Thai life and culture. From early-morning markets to late-night food stalls, the streets of Thailand are always buzzing with culinary activity. In this part, we'll guide you through the best street food spots and how to navigate the vibrant markets to make the most of your culinary adventure.

**Street Food**

Thailand is a street food paradise, and whether you're in the bustling metropolis of Bangkok or the more laid-back city of Chiang Mai, there's no shortage of delicious options to explore. Here are some of the best street food spots you shouldn't miss.

**Find the best spots**

**Yaowarat (Chinatown, Bangkok)**

Yaowarat, Bangkok's Chinatown, is a street food haven that comes alive after dark. The neon-lit streets are lined with food stalls offering everything from dim sum to seafood. Must-tries include oyster omelette (*hoi tod*), Chinese-style roast duck and crispy pork belly. For dessert, don't miss the mango sticky rice from the famous stall at the intersection of Yaowarat Road and Soi 9.

**Tip**: Arrive early to avoid the crowds and bring cash, as most vendors don't accept cards (although some will accept bank transfers via an account number or QR code).

**Khao San Road (Bangkok)**

Known as the backpacker hub, Khao San Road is not just about cheap hostels and lively bars; it's also a great place for street food. Here, you can find a variety of Thai classics like pad thai, spring rolls and satay skewers. For the adventurous, there are also exotic options like fried insects. The atmosphere is lively, and the food is both affordable and delicious.

**Tip**: Khao San Road is best visited in the evening when the street is closed to traffic and turns into a pedestrian paradise.

### Chatuchak Weekend Market (Bangkok)

Chatuchak Weekend Market is one of the largest markets in the world, and it's a treasure trove for street food lovers. With over 15,000 stalls, the market offers a little bit of everything. Be sure to try the coconut ice cream, grilled meats (*moo ping*) and papaya salad (*som tum*). Don't forget to wash it down with a refreshing Thai iced tea or local beer, like Singha or Tiger.

**Tip:** The market is massive, so wear comfortable shoes and come hungry. It's best to visit in the morning to avoid the midday heat.

### Sunday Night Market (Chiang Mai)

Chiang Mai's Sunday Night Market is a must-visit for street food enthusiasts. Stretching along Ratchadamnoen Road, the market offers a wide range of Northern Thai delicacies. Try the *khao soi*, a coconut curry noodle soup, and *sai oua*, a spicy Northern Thai sausage. For dessert, indulge in *kanom krok*, sweet coconut rice pancakes.

**Tip:** The market gets very crowded, so it's a good idea to arrive early. Also, be prepared to try small portions from multiple stalls to get a taste of everything.

### Warorot Market (Chiang Mai)

Warorot Market, also known as Kad Luang, is a bustling market near the Ping River. It's a great place to sample authentic Thai street food. Must-tries include *nam prik ong*, a Northern Thai chili dip served with vegetables, and *kao niew moo ping*, sticky rice with grilled pork skewers. The market also offers a variety of exotic fruits like mangosteen and rambutan.

**Tip:** This market is less touristy than others, providing a more authentic experience. It's best visited in the morning when the produce is freshest.

**Phuket Weekend Market (Phuket)**

Also known as Naka Market, the Phuket Weekend Market is a bustling hotspot for street food. Located just outside Phuket Town, this market offers a wide range of Southern Thai specialities. Try the massaman curry, *roti* (Thai-style pancakes) and grilled seafood. For dessert, don't miss the coconut pancakes.

**Tip:** The market is open only on weekends, so plan your visit accordingly. Don’t feel obliged to come when the sun is still up as it doesn’t close till 11pm.

**Old Town Night Market (Phuket)**

The Old Town Night Market in Phuket is a charming and vibrant spot to explore local street food. Located along Thalang Road, the market features a mix of traditional Thai dishes and international cuisine. Must-tries include hoy tod (oyster omelette), *mee hoon phuket* (stir-fried rice noodles), and *kanom jeen* (fermented rice noodles with curry).

**Tip:** Visit the market in the early evening to enjoy the beautiful colonial architecture of Old Town Phuket while sampling your food. Also, consider taking public transport as parking is limited.

**Cicada Market (Hua Hin)**

Cicada Market in Hua Hin is a unique blend of art, crafts and delicious street food. Open only on weekends, the market offers a variety of gourmet street food options. Try the Thai-style grilled squid, fresh seafood and pork skewers. The market also features live music and art installations, making it a great place to relax and enjoy your meal.

**Tip:** The market can get quite crowded, especially on weekends. Arrive early to find a good spot to sit and enjoy your food.

**Bang Lan Night Market (Ayutthaya)**

Located in the ancient city of Ayutthaya, Bang Lan Night Market is a fantastic spot to sample traditional Thai street food. The market is known for its variety of grilled meats, including moo ping (grilled pork skewers) and gai yang (grilled chicken). Other must-tries include *guay tiew reur* (boat noodles), a rich and flavourful noodle soup, and *khanom bueang*, Thai crispy pancakes.

**Tip:** Combine your visit to the market with a day of exploring Ayutthaya's historical sites. The market opens in the evening, making it a perfect end to a day of sightseeing.

## Markets

Navigating local markets in Thailand can be an exhilarating and tasty experience, but it can also be overwhelming if you don't know where to start. Here's a handy guide to help you make the most of your visit to Thai markets, ensuring you enjoy the best they have to offer without missing a beat.

### Navigating your way

**Plan your visit**

Before heading to any market, it's a good idea to plan your visit. Research the market's opening hours, peak times, and what types of goods are sold there. Some markets are morning affairs, while others come alive in the evening. Knowing when to go can make a big difference in your experience. For example, morning markets are ideal for fresh produce, while evening markets are perfect for street food and vibrant atmospheres.

**Bring cash**

Most street vendors and market stalls in Thailand operate on a cash-only basis. While larger markets in tourist areas may accept credit cards, it's safer to bring cash. Make sure to carry small denominations, as vendors may not always have change for larger bills.

**Wear comfortable clothing and shoes**

Markets can be sprawling and crowded, so it's important to dress comfortably. Wear lightweight, breathable clothing and comfortable shoes, as you'll likely be doing a lot of walking. A hat and sunscreen are also advisable when visiting a market during the day.

**Learn basic Thai phrases**

While many vendors in tourist areas speak some English, knowing a few basic Thai phrases can go a long way in enhancing your market experience. Simple phrases like 'hello' (*sawasdee*), 'thank you' (*khob khun*), and 'how much?' (*tao rai?*) can help you communicate more effectively with vendors.

**Haggling**

Haggling is a common practice in Thai markets, especially when buying non-food items like clothing, accessories, and souvenirs. Start by asking for the price and then politely negotiate. Keep it friendly and light-hearted; bargaining should be an enjoyable part of the experience.

**Try before you buy**

One of the best parts of visiting a Thai market is sampling the food. Many vendors offer free samples of their dishes, so don't hesitate to try a bite before committing to a full portion. This is a great way to discover new flavours and ensure you're getting something you'll enjoy.

**Be mindful of hygiene**

While street food in Thailand is generally safe, it's always wise to take some basic precautions. Choose stalls that are busy and have a high turnover of food, as this usually means the food is fresh. Avoid items that have been sitting out for a long time and opt for dishes cooked to order.

**Explore beyond the main areas**

While the main thoroughfares of a market are often bustling with activity, don't be afraid to venture into the smaller side streets and alleys. These areas can hold hidden gems and unique finds you might miss if you stick to the main paths.

**Respect local customs**

Thailand is known as the Land of Smiles, and showing respect and politeness is important. Always greet vendors with a smile and be courteous. If you decide not to buy something after inquiring about it, thank the vendor and move on.

**Enjoy the experience**

Visiting a Thai market is about more than just shopping; it's a cultural experience. Take in the sights, sounds, and smells, and enjoy the vibrant atmosphere. Engage with the vendors, learn about the products, and savour the unique flavours of Thai street food.

## Conclusion

Exploring Thailand's street food and markets is an adventure that offers a unique glimpse into the country's rich culinary and cultural heritage. From bustling spots like Bangkok's Yaowarat to the charming alleys of Chiang Mai's Warorot Market, each visit is a feast for the senses. By following these tips, you'll make the most of your market experiences, ensuring you savour the best of what Thai street food offers. So, put on your comfiest shoes, grab your wallet and dive into the vibrant and delicious world of Thai street food. Happy eating!

## COOKING AT HOME

One of the joys of living in Thailand as a digital nomad is the opportunity to immerse yourself in the local food culture, not just by eating out but also by cooking at home. Preparing your own meals allows you to explore Thai cuisine at your own pace, experiment with different ingredients and perhaps even develop your own twist on traditional recipes. Plus, it's a wonderful way to bring the flavours of Thailand into your daily life. In this section, we'll guide you through the essentials of shopping for groceries and ingredients in Thailand, ensuring you're well-equipped to start your culinary adventure.

### Shopping for Groceries and Ingredients

Shopping for groceries in Thailand can be an exciting experience, offering a mix of familiar and exotic ingredients. Whether you're a seasoned cook or a kitchen novice, here's everything you need to know to navigate Thai markets and supermarkets.

#### Supermarkets

Thailand has a variety of supermarkets that cater to both local and international tastes. Here are some of the most popular options.

- **Big C:** A hypermarket chain that offers a wide range of products, from fresh produce and meats to household items. Big C is a one-stop shop for all your grocery needs.
- **Lotus's (formerly Tesco Lotus):** Similar to Big C, Lotus's is a hypermarket chain with numerous locations

across Thailand. It offers a good selection of both local and imported products.

- **Makro:** Known for bulk buying, Makro is popular among small business owners and families. It's a great place to stock up on essentials, especially if you have a large household or entertain often.
- **Villa Market:** Catering more to expats, Villa Market offers a wide range of imported goods alongside local products. It's the place to go if you're looking for international brands and speciality items.
- **Tops Market:** Another supermarket that caters to locals and expats, Tops Market offers a good selection of fresh produce, meats, and international products.

**Popular local markets**

Thailand's local markets are vibrant hubs offering fresh produce, meats, seafood and unique local delicacies. Here are some must-visit markets across the country.

- **Khlong Toei Market (Bangkok):** One of the largest fresh markets in Bangkok, Khlong Toei Market offers various fruits, vegetables, meats and seafood. It's a bustling, vibrant place that provides a true taste of local life.
- **Or Tor Kor Market (Bangkok):** Located near the famous Chatuchak Market, Or Tor Kor is known for its high-quality produce and artisanal products. It's slightly pricier but offers some of the best ingredients you can find.
- **Warorot Market (Chiang Mai):** Also known as Kad Luang, this market is a favourite among locals for its fresh produce and traditional Northern Thai ingredients. It's a great place to find unique items like *sai oua* (Northern Thai sausage) and *nam prik* (chilli paste).

- **Muang Mai Market (Chiang Mai):** Another must-visit in Chiang Mai, Muang Mai Market is a wholesale market that offers fresh produce at bargain prices. It's a bustling market where you can find everything from tropical fruits and vegetables to fresh herbs and spices. It's particularly known for its impressive selection of fresh fruits.
- **Banzaan Fresh Market (Phuket):** Located near Patong Beach, Banzaan Fresh Market is a modern, clean market that offers a wide range of fresh produce, seafood and meats. The market also has a food court where you can cook your seafood on the spot. It's a convenient place for tourists to experience a local market without straying too far from the main attractions.
- **Phuket Town Central Market (Phuket):** Another gem in Phuket, this market offers a variety of fresh produce, meats and seafood. It's a bustling market where locals shop, providing an authentic experience. You'll find everything from tropical fruits and local vegetables to various seafood. It's an excellent place to immerse yourself in the local culture and pick up fresh ingredients for your Thai dishes.
- **Talad Kaset Market (Phitsanulok):** This local market in Phitsanulok offers a variety of fresh produce, meats and seafood. It's a bustling market where you can find everything from tropical fruits and vegetables to fresh herbs and spices. It's a great place to experience local life and shop for fresh ingredients.
- **Krabi Town Walking Street (Krabi):** This market is more than just a place to shop; it's a social hub where locals and tourists mingle. The market offers a variety of street food, fresh produce and handmade crafts. It's open on weekends and is a great place to experience the local culture and cuisine.

- **Talat Sao (Udon Thani):** One of the largest markets in Udon Thani, Talat Sao is a bustling hub for fresh produce, meats and seafood. It's a great place to find local specialities and experience the vibrant market culture of Northeastern Thailand.
- **Rong Kluea Market (Sa Kaeo):** Located near the Cambodian border, Rong Kluea Market offers a unique shopping experience with a mix of Thai and Cambodian goods. The market is known for its fresh produce, seafood and a variety of other products. It's a great place to find unique ingredients and experience the local culture.
- **Talat Ton Phayom (Chiang Rai):** This local market in Chiang Rai offers a variety of fresh produce, meats and seafood. It's a bustling market where you can find everything from tropical fruits and vegetables to fresh herbs and spices. It's a great place to experience local life and shop for fresh ingredients.

**Local market shopping tips**

For a more authentic shopping experience, local markets are the way to go. These markets offer fresh, seasonal produce and a variety of other ingredients at reasonable prices. Here are some tips for shopping at local markets.

- **Early bird advantage:** Local markets are usually busiest in the early morning. Arriving early ensures you get the freshest produce and the best selection. Plus, it's a great way to beat the heat and the crowds.
- **Know your seasons:** Thailand's tropical climate means different fruits and vegetables are available at different times of the year. Familiarise yourself with what's in season to get the best quality and prices.

- **Bring your own bags:** Many local markets are trying to reduce plastic waste, so bringing your own reusable bags is both eco-friendly and practical.
- **Learn basic Thai phrases:** Knowing a few phrases like 'how much?' (*tao rai?*), 'I want' (*ao*), and 'I don't need a bag' (*mai ao thung*) can make your shopping experience smoother and more enjoyable.
- **Haggling and bulk buying:** Haggling is common in local markets, especially for non-food items. Start by offering about 50-70% of the asking price and negotiate from there. When buying produce, meats, or seafood in bulk, you can often get a discount. Politely ask for a better price, and always keep the negotiation friendly and respectful.

**Specialty stores**

If you're looking for specific ingredients or international products, speciality stores are your best bet. These stores often carry items that are hard to find in regular supermarkets or local markets.

- **Gourmet Market (nationwide):** Located in upscale shopping centres like Siam Paragon and CentralFestival, Gourmet Market offers a wide range of imported goods, premium meats and gourmet items. It's a bit pricier but perfect for finding those hard-to-get ingredients.
- **Villa Market (nationwide):** As mentioned earlier, Villa Market caters to expats and offers a variety of imported goods. It's a reliable place to find international brands and speciality products that you might not find elsewhere.
- **Sunshine Market (Bangkok):** Sunshine Market is a health-conscious grocery store offering organic produce, gluten-free products and a variety of healthy foods. It's a

great spot for those looking to maintain a healthy lifestyle while in Thailand.

- **Rimping Supermarket (Chiang Mai):** Known for its wide selection of international products and organic produce, Rimping Supermarket is a favourite among expats in Chiang Mai. It's a great place to find Western cooking essentials and speciality items.
- **Panama Market (Phuket):** Panama Market is a well-stocked store offering a variety of imported goods, fresh produce, and speciality items at multiple locations. It's popular among expats (especially Russians) and tourists looking for international products and quality ingredients.
- **BritiShop (online):** BritiShop is an online store specialising in British food products. From baked beans to Marmite, it's the go-to place for British and Aussie expats missing a taste of home.
- **iHerb (online):** This is an online platform offering various herbs, spices, supplements and unique cooking ingredients. It's perfect for those looking to add a healthy twist to their cooking.
- **Ninelife (online):** Ninelife is an e-commerce site that offers a variety of international products, including groceries, health foods and speciality items. It's a convenient option for finding unique ingredients from around the world.

**Online grocery shopping**

For those who prefer the convenience of online shopping, several platforms offer grocery delivery services in Thailand. Here's a detailed look at the options available:

**7 Delivery**

7-Eleven's delivery service brings convenience store items directly to your door.

- Minimum spend for free delivery: **THB 100**
- Delivery fee: **THB 10-20** if below the minimum spend
- Delivery lead time: Within 1-2 hours

**Bangkok Grocery**

An online store offering a variety of groceries, including imported items and speciality foods.

- Minimum spend for free delivery: **THB 1,000**
- Delivery fee: **THB 60** if below the minimum spend
- Delivery lead time: Within 24 hours

**Big C**

Big C's online service offers a comprehensive selection of groceries, household items, and more.

- Minimum spend for free delivery: **THB 1,500**
- Delivery fee: **THB 59-99** if below the minimum spend
- Delivery lead time: Same-day delivery available

**Foodpanda**

Best known for food delivery, Foodpanda also offers a grocery delivery service from local supermarkets and convenience stores.

- Minimum spend for free delivery: Varies by store
- Delivery fee: Varies by store
- Delivery lead time: Same-day delivery available

**Gourmet Market**

The online platform for Gourmet Market is available in upscale shopping centres like Siam Paragon and CentralFestival.

- Minimum spend for free delivery: **THB 1,500**
- Delivery fee: **THB 100** if below the minimum spend
- Delivery lead time: Within 24 hours

**Grab**

Known primarily for ride-hailing, Grab also offers grocery delivery services from local supermarkets and stores.

- Minimum spend for free delivery: Varies by store
- Delivery fee: Varies by store
- Delivery lead time: Same-day delivery available

**HappyFresh**

An online grocery delivery service that partners with local supermarkets like Big C, Lotus's, and Villa Market. It offers a wide range of products and delivers straight to your door.

- Minimum spend for free delivery: Typically **THB 1,500**
- Delivery fee: Starts at **THB 50** if below the minimum spend
- Delivery lead time: Same-day delivery available

**Lazada and Shopee**

These popular e-commerce platforms also offer a variety of grocery items, including non-perishable goods, speciality ingredients, and imported products.

- Minimum spend for free delivery: Varies by seller
- Delivery fee: Varies by seller
- Delivery lead time: Typically 2-5 days

**Lotus's**

Lotus's online service offers a wide range of products, including fresh produce, household items, and more.

- Minimum spend for free delivery: **THB 1,200**
- Delivery fee: **THB 60-120** if below the minimum spend
- Delivery lead time: Within 24 hours

**Makro**

Makro's online platform is perfect for bulk buying with a wide range of products.

- Minimum spend for free delivery: **THB 2,000**
- Delivery fee: **THB 100** if below the minimum spend
- Delivery lead time: Within 24 hours

**Tops**

The online platform for Tops Market offers the same selection of products with the added convenience of home delivery.

- Minimum spend for free delivery: **THB 888**
- Delivery fee: **THB 60-80** if below the minimum spend
- Delivery lead time: Same-day delivery available

**We Fresh**

We Fresh offers a wide range of fresh produce, meats, and seafood delivered to your door.

- Minimum spend for free delivery: **THB 1,500**
- Delivery fee: **THB 50-100** if below the minimum spend
- Delivery lead time: Same-day delivery available

## Taking a Thai Cooking Class

Cooking at home is a fantastic way to immerse yourself in Thai culture, and what better way to learn than by taking a Thai cooking class? Whether you're a novice in the kitchen or an experienced home cook, a cooking class can help you master the art of Thai cuisine, explore new ingredients and understand the cultural significance behind each dish. Here's everything you need to know about taking a Thai cooking class in Thailand.

## Why learn to cook Thai food?

Thai cooking classes offer more than just recipes; they provide a holistic experience that combines learning, tasting, and cultural immersion. Here are some reasons why you should consider taking one.

- **Hands-on experience:** Cooking classes provide practical, hands-on experience, allowing you to learn by doing. This is especially useful for mastering techniques specific to Thai cuisine, like balancing flavours and using unique ingredients.
- **Cultural insights:** Many cooking classes include a market tour, where you'll learn about local ingredients and their cultural significance. This adds depth to your cooking and helps you understand the broader context of Thai food.
- **Confidence in the kitchen:** Whether making a spicy tom yum soup or a delicate mango sticky rice, a cooking class will boost your confidence in preparing these dishes at home.
- **Meeting fellow food enthusiasts:** Cooking classes are a great way to meet like-minded travellers and locals who share your passion for food.

## Choosing the right class and what to expect

Thailand offers numerous cooking schools, each providing unique experiences. Here's how to choose the right one.

### Selection

- **Location:** Cooking classes are available in major cities and tourist destinations like Bangkok, Chiang Mai, Phuket, and Krabi. Many offer free pick-up and drop-off from your accommodation.

- **Class size:** Smaller classes offer more personalised attention. Check the maximum number of participants before booking.
- **Course content:** Classes vary from street food specialities to comprehensive menus covering appetisers, main courses, and desserts. Choose one that focuses on the dishes you're most interested in learning.
- **Duration and cost:** Classes range from a few hours to full-day experiences. Shorter classes typically cost around **THB 1,000-2,000**, while full-day courses cost upwards of **THB 3,000**. Consider your schedule and budget.

**Expectations**

- **Market tour:** Many classes start with a visit to a local market to learn about Thai ingredients, select the freshest produce and sample local delicacies.
- **Preparation:** Back at the school, the instructor guides you through ingredient preparation, explaining the significance of each one.
- **Cooking:** Follow step-by-step instructions at cooking stations, with hands-on tips and corrections from the instructor.
- **Tasting:** Enjoy the dishes you've prepared in a communal meal, often accompanied by rice and traditional sides.
- **Recipes:** Many schools provide recipe booklets to help you recreate the dishes at home, including tips and variations for adapting them to your taste and local ingredients.

## Recommended cooking schools

Here are some highly recommended cooking schools across the country to consider.

**Bangkok**

- **Baipai Thai Cooking School:** Located in a beautiful garden setting, Baipai offers half-day courses that include a market tour and hands-on cooking experience. Courses typically cost around **THB 2,200**.
- **Sompong Thai Cooking School:** Known for its friendly instructors and well-organised classes, Sompong offers half-day classes focusing on traditional Thai dishes. Prices start at **THB 1,200**.

**Chiang Mai**

- **Chiang Mai Thai Cookery School:** One of the oldest cooking schools in the city, it offers a variety of courses, including full-day experiences. Prices start at around **THB 1,500** for a half-day course.
- **Thai Farm Cooking School:** Located on an organic farm, this school offers immersive full-day courses that include a farm tour and hands-on cooking. Prices are around **THB 1,300**.

**Phuket**

- **Phuket Thai Cooking Academy:** Offering classes in a serene setting, this school provides half-day and full-day courses focusing on Southern Thai cuisine. Prices start at **THB 2,500**.
- **Blue Elephant Cooking School:** Part of the renowned Blue Elephant restaurant, this school offers premium cooking classes in a luxurious setting. Prices start at **THB 3,000** for a half-day course.

**Krabi**

- **Ya's Cookery School:** Known for its friendly atmosphere and stunning views, Ya's offers half-day and full-day courses focusing on Thai favourites. Prices start at **THB 1,500**.

- **Smart Cook Thai Cookery School:** Offers a variety of courses, including full-day experiences. Located in a beautiful setting, prices start at **THB 1,200**.

**Socially conscious cooking schools**

For those who want to combine their love of cooking with social responsibility, here are three top-tier cooking schools across Thailand that focus on giving back to the community.

- **Cooking with Poo and Friends (Bangkok):** This unique cooking school is run by Khun Poo, a resident of Klong Toey, one of Bangkok's largest slums. The school offers hands-on cooking classes that teach traditional Thai recipes while providing employment opportunities and supporting the local community. Courses typically cost around **THB 1,500**.
- **Thai Akha Kitchen (Chiang Mai):** This cooking school not only teaches traditional Thai and Akha Hill Tribe recipes but also supports the local Akha community. The classes often include a market tour and hands-on cooking experience, allowing participants to learn about the unique ingredients and cooking techniques of the Akha people. Prices start at around **THB 1,200**.
- **Community-Based Tourism Cooking Classes (Phuket):** Located in various villages across Phuket, these cooking classes are part of a broader community-based tourism initiative. Participants learn to cook traditional Thai dishes using locally sourced ingredients, and the proceeds support local community projects and sustainable tourism efforts. Prices vary depending on the village and course length but typically start around **THB 1,800**.

**Conclusion**

Cooking at home in Thailand is an adventure that allows you to immerse yourself in the rich culinary culture of the country. From shopping for fresh, seasonal ingredients at local markets and speciality stores to taking a Thai cooking class that offers hands-on experience and cultural insights, you'll find countless ways to enhance your culinary skills. Whether sourcing unique ingredients from a gourmet market or learning traditional recipes at a socially conscious cooking school, the opportunities are endless. Embrace the flavours, enjoy the process and bring the vibrant essence of Thai cuisine into your home kitchen. Happy cooking!

# DIETARY RESTRICTIONS

Thailand is a food lover's paradise, but navigating its culinary landscape with dietary restrictions can be challenging. Whether you're vegetarian, vegan, gluten-free or dealing with food allergies, understanding how to communicate your needs and knowing where to find appropriate options is crucial. This section will guide you through the essentials of managing dietary restrictions in Thailand, starting with vegetarian and vegan options.

## Vegetarian and Vegan Diet

Thailand's rich culinary tradition naturally accommodates vegetarians and vegans, thanks to its abundant use of fresh vegetables, herbs and spices. However, the prevalence of fish sauce, shrimp paste and meat-based broths in many dishes can pose challenges. Here's how to enjoy Thai cuisine while following a vegetarian or vegan diet.

### Thai cuisine for vegetarians and vegans

Thai food is known for its balanced flavours – sweet, sour, salty, bitter and spicy. While many dishes inherently feature vegetables, they often include animal products like fish sauce or shrimp paste. Understanding the key ingredients and how to request modifications is essential for maintaining a vegetarian or vegan diet. Many Thai dishes can be adapted by simply substituting soy sauce for fish sauce or requesting the dish to be made without shrimp paste. Additionally, coconut milk is often used as a base for curries, making it easier to find vegan-friendly options.

Here are some traditional Thai dishes that can be easily adapted for vegetarians and vegans.

- ***Pad pak ruam mit*:** Stir-fried mixed vegetables. Ensure it's made with soy sauce instead of fish sauce.
- ***Som tum*:** Green papaya salad. Ask for it without fish sauce or dried shrimp.
- ***Gaeng jay*:** Vegetarian curry. Ensure it's made with coconut milk and no fish sauce or shrimp paste.
- ***Pad Thai jay*:** Vegetarian pad Thai. Request no egg and fish sauce.
- ***Tom yum hed*:** Spicy mushroom soup. Ensure it's made without fish sauce or meat broth.

**Key phrases to know**

Learning a few basic Thai phrases can help ensure your dietary needs are met.

- **Vegetarian:** 'Chan gin jay' (for women) / 'Pom gin jay' (for men) – This indicates you follow a vegetarian diet, often understood to exclude meat, fish, and sometimes eggs and dairy.
- **Vegan:** 'Chan mai gin sat, kai, nom' (for women) / 'Pom mai gin sat, kai, nom' (for men) – This means you don't eat meat, eggs or dairy.
- **No fish sauce:** 'Mai sai nam pla'.
- **No shrimp paste:** 'Mai sai gapi'.

- **No meat:** 'Mai sai neua-sat'.

**Navigating street food and local markets**

While dedicated vegetarian and vegan restaurants are a safe bet, don't miss out on the vibrant street food scene and local markets. Here are some tips for finding vegetarian and vegan options.

- **Look for jay symbols:** The Thai word 'jay' (เจ) signifies vegetarian food and is often used to label dishes free of meat, fish sauce, and certain pungent vegetables like garlic and onions.
- **Visit vegetarian food stalls:** Many markets have dedicated vegetarian vendors, especially during the annual Vegetarian Festival in October.
- **Ask for customisations:** Don't hesitate to ask vendors to prepare dishes without meat or animal products. Most are happy to accommodate if you communicate your needs clearly.
- **Explore fruit stalls:** Thailand's markets are brimming with fresh, exotic fruits like mangoes, papayas, and dragon fruits. These are naturally vegan and can be a great snack or dessert.

**Resources and apps**

Several resources and apps can make navigating Thailand's vegetarian and vegan food scene easier.

- **Google Translate:** Handy for translating ingredient lists and menus. While not perfect, it can be a useful tool for navigating dietary restrictions.
- **HappyCow:** A popular app for finding vegetarian and vegan restaurants worldwide. It provides user reviews, photos and maps.

- **Vegan Passport:** A multilingual phrasebook that helps you communicate your dietary needs in different languages, including Thai.
- **Vegan-Friendly Destinations in Thailand:** Offers recommendations for vegetarian and vegan dining locations (cities) in Thailand.
- **Vegetarian and Vegan Guide to Thailand:** This comprehensive guide provides info on local dishes, ingredient substitutes, and essential phrases to help you enjoy your culinary adventure.

## Dealing with Food Allergies

Travelling with food allergies can be daunting, but with a bit of preparation, you can safely enjoy the rich and diverse cuisine Thailand has to offer. Whether you have a peanut allergy, shellfish sensitivity or lactose intolerance, here's how to navigate Thai food while keeping your health in check.

### Common food allergens in Thai cuisine

Thai cuisine uses a wide range of ingredients, some of which are common allergens. Here are some of the most prevalent allergens and where you might encounter them.

- **Peanuts:** Often found in dishes like pad thai, som tum (papaya salad), and various dipping sauces.
- **Shellfish:** Commonly used in soups like tom yum goong, curries and stir-fries.
- **Fish sauce and shrimp paste:** Ubiquitous in Thai cooking, found in curries, stir-fries, soups and salads.
- **Soy:** Present in tofu dishes and soy sauce-based stir-fries.
- **Gluten:** Found in soy sauce, wheat noodles and some desserts.
- **Dairy:** Less common in traditional Thai cuisine but can be found in some fusion dishes and desserts.

### Key phrases to know

Learning a few essential Thai phrases can help you communicate your allergies effectively.

- **I'm allergic to peanuts:** *Chan pae tua lisong* (for women) / *Pom pae tua lisong* (for men)
- **I'm allergic to shellfish:** *Chan pae ahan talay* (for women) / *Pom pae ahan talay* (for men)
- **I'm allergic to fish sauce:** *Chan pae nam pla* (for women) / *Pom pae nam pla* (for men)
- **I'm allergic to soy:** *Chan pae tua leung* (for women) / *Pom pae tua leung* (for men)
- **I'm allergic to gluten:** *Chan pae gluten* (for women) / *Pom pae gluten* (for men)
- **I'm allergic to dairy:** *Chan pae nom* (for women) / *Pom pae nom* (for men)

### Tips for managing food allergies in Thailand

Having a food allergy doesn't need to derail your travel plans; you just need to be prepared.

- **Carry an allergy card:** Write your allergies in Thai on a card that you can show to restaurant staff and street food vendors. This can be a lifesaver in ensuring your food is prepared safely.
- **Research ahead:** Before dining out, research restaurants are known for accommodating food allergies. Apps like **HappyCow** and websites like TripAdvisor can provide useful reviews and recommendations.
- **Choose fresh, simple dishes:** Opt for dishes with easily identifiable ingredients. Fresh salads, stir-fried vegetables, and steamed rice are generally safer choices.
- **Communicate clearly:** When ordering food, state your allergies and ask the staff to confirm that your dish will

be prepared without the allergen. Don't hesitate to ask questions about ingredients and preparation methods.

- **Be cautious with street food:** While street food is a major attraction in Thailand, it can be riskier for those with food allergies due to cross-contamination and the use of common ingredients like fish sauce and peanuts. If you're unsure, it's better to avoid it.
- **Pack emergency medication:** Always carry antihistamines and, if necessary, an epinephrine auto-injector (EpiPen). Make sure your travel companions know how to use it in case of an emergency.
- **Stay in accommodation with a kitchen:** If possible, choose accommodation with kitchen facilities to prepare your own meals. This gives you complete control over what you're eating and ensures your food is safe.

**Resources and apps**

Several resources and apps can make managing food allergies in Thailand easier.

- **AllergyEats:** A user-driven app that helps you find restaurants that accommodate food allergies.
- **Food Allergy Translation App:** This app by Ask Allergies translates your food allergies into different languages, including Thai, and provides a printable allergy card.
- **HappyCow:** A popular app for finding vegetarian, vegan, and allergy-friendly restaurants worldwide. It provides user reviews, photos, and maps.
- **Spokin:** An allergy management app that offers a variety of features, including restaurant reviews, product recommendations, and travel tips specifically for people with food allergies.

**Conclusion**

Navigating dietary restrictions in Thailand, whether you're vegetarian, vegan or dealing with food allergies, is entirely possible with a bit of preparation and knowledge. From learning key phrases to researching allergy-friendly restaurants, these tips will help you enjoy the rich and diverse flavours of Thai cuisine safely. Whether you're indulging in a vegan pad Thai or ensuring your meal is peanut-free, you can savour the culinary delights Thailand has to offer with confidence and peace of mind. Embrace the experience, stay informed, and enjoy the vibrant world of Thai food while keeping your dietary needs in check.

**Stay tuned . . .**

Get ready to dive into the heart of Thailand's vibrant social scene in the next chapter: **Community & Social Life**. Discover how to build your tribe and make lasting friendships while living as a digital nomad in the Land of Smiles. We'll explore popular hangouts and events, including co-working spaces, cafes, and meetups where you can connect with fellow nomads. Learn about joining local communities through volunteering and cultural exchange programmes, such as homestay and language immersion. This chapter will also guide you through balancing social life and work, ensuring you make the most of your time in Thailand.

Stay tuned for tips on navigating social media groups and specialised forums to find your community, as well as attending must-see events like traditional festivals and technology seminars. We'll discuss the importance of connecting with loved ones back home through video calls, social media, and care packages. Additionally, you'll find advice on making long-term friendships, engaging in local activities, and connecting with both locals and expats. Don't miss **Chapter 9** – your ultimate guide to building a fulfilling social life in Thailand!

# Chapter 9: Community & Social Life

## Building Your Tribe: Making Friends in Thailand

Living in a new country can be daunting, but building a supportive community can make all the difference. Thailand's vibrant expat and digital nomad scene offers countless opportunities to connect with like-minded individuals and immerse yourself in the local culture. Whether you're joining a co-working space, participating in social media groups, or attending local festivals, there are endless ways to meet new people and form lasting friendships. Volunteering and language exchange programmes provide unique avenues to engage with the local community. In this chapter, we'll explore the best strategies for integrating into Thai society, making friends, and creating a fulfilling social life.

## MEETING OTHER DIGITAL NOMADS

One of the most rewarding aspects of being a digital nomad is the chance to meet like-minded people from all over the world. Thailand, with its welcoming atmosphere and thriving expat community, offers countless opportunities to connect with fellow nomads. Whether you're in bustling Bangkok, laid-back Chiang Mai, or the beachy paradise of Phuket, you'll find plenty of popular hangouts and events to help you build your network. Let's dive into some of the best spots and gatherings where digital nomads unite.

### Popular Hangouts and Events

#### Co-working spaces

Co-working spaces are the heart of the digital nomad community in Thailand. These spaces not only provide a productive work environment but also serve as social hubs where you can meet other remote workers. Here are some of the top co-working spaces across the country.

**Bangkok**

In the capital city, you'll find a wide variety of co-working spaces catering to different needs and budgets. Spaces like **The Hive, UnionSPACE, The Work Loft**, and **WOLF Co-working Space** are popular among digital nomads for their excellent facilities and vibrant communities. Many of these spaces host regular networking events, workshops, and social gatherings, making it easy to meet new people and form connections.

**Chiang Mai**

Often referred to as the digital nomad capital of the world, Chiang Mai boasts numerous co-working spaces that attract remote workers from all corners of the globe. **Punspace**, with its multiple locations, is a favourite among nomads for its relaxed atmosphere and strong community vibe. Other notable spaces include **Greenhouse**, **Yellow Co-working**, and **Starwork**, all offering great amenities and frequent events.

**Phuket**

If the beach is more your scene, Phuket has several co-working spaces that cater to digital nomads. **Garage Society, Let's Work, Homa**, and **Regus Phuket** are popular options that provide a perfect blend of work and leisure. These spaces often organise meetups, networking sessions, and even beach parties, allowing you to connect with fellow nomads in a more relaxed setting.

*For more in-depth knowledge of co-working spaces, check out **Chapter 3: Finance & Budgeting**.

**Cafés and coffee shops**

Thailand's café culture is thriving, and many digital nomads prefer working in cosy coffee shops. These spots are not only great for getting work done but also excellent for meeting new people. Here are some top picks:

**Bangkok**

The city is teeming with trendy cafés that cater to remote workers. **Rocket Coffee Bar**, **Casa Lapin**, **Roots Coffee**, and **The COMMONS** are just a few examples of popular spots where you can sip on a delicious brew while working on your laptop. These cafés often have communal tables, making it easy to converse with fellow nomads.

**Chiang Mai**

With its laid-back vibe, Chiang Mai is home to numerous cafés that attract digital nomads. **Ristr8to**, **The Barisotel by The Baristro**, **Akha Ama Coffee**, and **Nine One Coffee** are well-loved for their excellent coffee and comfortable workspaces. Many of these cafés have strong Wi-Fi and plenty of power outlets, making them ideal for a productive workday.

**Phuket**

In Phuket, you'll find a mix of beachfront cafés and urban spots perfect for remote work. **Coffee Lab**, **Project Artisan**, **Umai Teahouse**, and **Bookhemian** are popular among digital nomads. These cafés offer a relaxed atmosphere where you can work, network, and enjoy the island's stunning scenery.

**Meetups and social events**

Attending meetups and social events is one of the best ways to expand your network and make new friends. Thailand has a vibrant calendar of events catering to digital nomads, from casual hangouts to professional networking sessions. Here are some to keep an eye out for:

### Bangkok

The city hosts numerous events for digital nomads, including networking nights, workshops, and social gatherings. **Bangkok Entrepreneurs** and **BNOW.org** are two groups that organise regular meetups and events focused on business and entrepreneurship. Additionally, co-working spaces like **The Hive** and **WeWork** often host their own events, providing ample opportunities to meet fellow nomads.

### Chiang Mai

Known for its strong digital nomad community, Chiang Mai offers a wide range of events designed to help remote workers connect. **Nomad Coffee Club** is a popular weekly meetup where nomads gather to discuss their projects, share tips, and network. Other notable events include **Digital Nomad Meetups**, and **Chiang Mai Expats Club** events, providing social, cultural and education opportunities.

### Phuket

While the digital nomad scene in Phuket may not be as large as in Bangkok or Chiang Mai, there are still plenty of events to help you connect with others. **Phuket Startup Businesses** and **Phuket Expats Club** organise regular meetups and social events, providing a platform for networking and making new friends. Many co-working spaces in Phuket, such as **Garage Society** and **Hatch Co-working Space**, also host their own events, so be sure to check their calendars.

### Conferences and summits

For those looking to dive deeper into the digital nomad lifestyle and gain valuable insights, attending conferences and summits is a great option. Thailand hosts several major events that attract nomads from around the world.

- **D.NX Conference:** While not exclusive to Thailand, the D.NX Conference often has events in major cities like Bangkok. This international conference focuses on digital nomadism, remote work, and entrepreneurship. It's a fantastic opportunity to learn from industry leaders, attend workshops, and network with fellow digital nomads.
- **Dragonfly:** Dragonfly is an annual event held in Bangkok that focuses on mindfulness, creativity, and personal development. It includes workshops, talks, and activities designed to inspire and connect like-minded individuals, fostering a sense of community and growth.
- **Nomad Summit:** Held annually in Chiang Mai, the Nomad Summit is one of the largest gatherings of digital nomads. The event features a lineup of expert speakers, workshops, and networking sessions, all designed to help you grow your business and connect with like-minded individuals. With hundreds of attendees, the Nomad Summit is an excellent opportunity to expand your network and gain inspiration from successful nomads.
- **Overseas offerings:** If you're keen to explore events beyond Thailand, nearby cities like Singapore, Kuala Lumpur, and Hong Kong host notable conferences and meetups. Events such as **RISE Conference** in Hong Kong, **Tech in Asia** in Singapore, and **Wild Digital** in Kuala Lumpur are excellent opportunities for networking and professional development
- **Techsauce Global Summit:** Held in Bangkok, this is one of Asia's leading technology conferences. It gathers entrepreneurs, investors, and professionals from various industries to discuss the latest trends in technology and innovation. The summit features keynote speakers, panel discussions, and networking sessions.

- **Thailand Wellness Summit:** This summit focuses on holistic health, wellness, and sustainable living. Held in various locations across Thailand, it offers workshops, presentations, and networking opportunities with wellness experts and enthusiasts.

## Online Communities and Forums

In today's digital age, online communities and forums are invaluable resources for digital nomads looking to connect, share experiences, and find support. Whether you're seeking advice on visa applications, looking for the best co-working spaces, or simply wanting to chat with like-minded individuals, there's a thriving online community ready to welcome you. Let's explore some of the best online platforms where digital nomads in Thailand gather.

### Social media groups

Social media platforms, particularly Facebook, are teeming with groups dedicated to digital nomads in Thailand. These groups are excellent for networking, gathering information, and staying up-to-date with events and opportunities.

- **Bangkok Digital Nomads:** For those based in the bustling capital, Bangkok Digital Nomads is a fantastic resource. The group covers many topics, including visa advice, accommodation recommendations, and social events. It's also a great place to find job opportunities, freelance gigs, and business partnerships.
- **Chiang Mai Digital Nomads:** If you're heading to Chiang Mai, this group is a must-join. Known as the digital nomad capital of the world, Chiang Mai has a vibrant online community where you can find information on the best co-working spaces, upcoming events, and local tips. The group also serves as a

platform for organising meetups and social activities, helping you quickly integrate into the local scene.

- **Digital Nomads in Thailand:** This Facebook group is one of the largest and most active communities for digital nomads in Thailand. With thousands of members, it's a great place to ask questions, share experiences, and find recommendations on everything from accommodation to local activities. Group members are generally very responsive and helpful, making it an excellent resource for newcomers and seasoned nomads.
- **Thailand Expats:** While not exclusively for digital nomads, this group offers valuable insights and advice for anyone living in Thailand. From navigating Thai bureaucracy to finding the best restaurants, the group's members cover a broad spectrum of topics. It's particularly useful for understanding the nuances of Thai culture and integrating into the local community.

**Specialised forums and websites**

Beyond social media, several specialised forums and websites cater specifically to digital nomads and expats in Thailand. These platforms often provide more in-depth information and facilitate more focused discussions.

- **Expat.com:** Expat.com is a global platform with a dedicated section for Thailand. The website offers forums, blogs, and guides covering various aspects of expat life, including visas, housing, and healthcare. The community is very active, and the forums are a great place to seek advice and share your experiences.
- **Nomads.com:** Nomas.com is a comprehensive resource for digital nomads, offering detailed information on cities worldwide, including several in Thailand. The platform includes data on the cost of living, internet speed, safety, and quality of life, making it an excellent

tool for planning your move. Nomad List also has an active community forum where you can ask questions, share experiences, and connect with other nomads.

- **Reddit:** The **r/digitalnomad** and **r/Thailand** subreddits are valuable resources for digital nomads in Thailand. These forums cover various topics, from visa issues and accommodation tips to personal stories and travel advice. The Reddit community is known for its candid and often humorous discussions, making it a fun and informative place to gather insights.
- **Thaivisa:** Thaivisa is one of the oldest and most comprehensive resources for expats in Thailand. The website includes news articles, forums, and guides on everything from legal issues to lifestyle tips. The forums are particularly valuable for detailed discussions on topics like visa extensions, work permits, and property rentals.
- **WiFI Tribe:** WiFi Tribe is a community of remote professionals who travel the world together while working online. They offer a unique way to experience different countries, including Thailand, by organising month-long co-living and co-working arrangements. Members benefit from a supportive network, shared experiences, and professional growth opportunities.

**Real-time interactions and virtual events**

From immediate and casual interactions to structured forums and webinars, the digital landscape offers numerous ways for digital nomads to connect, learn, and grow. Real-time platforms like **Slack** and **Discord** facilitate quick and engaging conversations, while virtual events and webinars provide in-depth learning opportunities and networking with industry experts. Here are some of the best platforms and events to help you stay connected and informed.

- **Digital Nomad Slack:** The Digital Nomad Slack group is a global community with channels dedicated to various regions, including Thailand. Within the Thailand channel, you can find discussions on everything from the best co-working spaces to local events. The Slack platform's real-time nature makes it easy to have fluid and engaging conversations.
- **Nomad City online conference:** Nomad City, a well-known conference for digital nomads, has expanded to include virtual editions. The online events feature workshops, networking opportunities, and talks from influential speakers, focusing on remote work trends and best practices.
- **Nomad Soulmates Discord:** Nomad Soulmates is a community focused on helping digital nomads build meaningful connections, professionally and personally. The Discord server includes channels for different regions, including Thailand. It's a great place to meet new people, find travel buddies, and explore potential romantic connections.
- **Remote work Meetups:** Platforms like Meetup.com often host virtual events focused on remote work and digital nomadism. These meetups cover a wide range of topics, from productivity tips to mental health, and offer opportunities to network with other remote workers.
- **Virtual Nomad Summit:** The Nomad Summit, a major event for digital nomads, has started offering virtual editions. These online conferences feature keynote speakers, panel discussions, and networking sessions, all accessible from the comfort of your home. It's a fantastic way to connect with the broader nomad community and gain valuable knowledge.
- **WebWork Travel webinars:** WebWork Travel hosts regular webinars on various topics related to remote

work and digital nomadism. These sessions often feature industry experts and provide practical advice on improving productivity, managing remote teams, and maintaining work-life balance.

**Conclusion**

Thailand offers an abundance of opportunities for digital nomads to connect and build lasting relationships, both in-person and online. From bustling co-working spaces and charming cafés to vibrant meetups and major conferences, the country is teeming with popular hangouts and events that foster community. Additionally, online platforms like Facebook Groups, specialised forums, and real-time chat communities provide invaluable support and networking opportunities. By engaging with these diverse avenues, you'll not only enhance your social life but also gain valuable insights and support, making your digital nomad journey in Thailand both rewarding and unforgettable. Jump in, connect, and thrive!

## FINDING YOUR TRIBE

While connecting with fellow digital nomads is essential, immersing yourself in the local Thai community can be an equally enriching experience. Joining local communities allows you to gain a deeper understanding of Thai culture, make meaningful contributions, and build lasting relationships with locals and expats alike. Whether through volunteering, joining clubs, or participating in cultural exchange programmes, becoming part of the local fabric can make your stay in Thailand truly special. In this section, we'll explore various ways to integrate into local communities, starting with volunteering and local clubs.

## Getting Out There

### Volunteering opportunities

Volunteering can be a rewarding way to give back to the local community while building connections with other volunteers and locals. Thailand offers many volunteer opportunities across various sectors, including education, animal welfare, community development, environmental conservation, sports coaching, and beach cleanups.

- **Animal welfare:** Thailand has numerous animal sanctuaries and rescue centres that rely on volunteers for support. The **Elephant Nature Park** in Chiang Mai is a well-known sanctuary that rescues and rehabilitates elephants. Volunteers can help with feeding, bathing, and caring for these majestic animals. Similarly, the **Soi Dog Foundation** in Phuket focuses on rescuing and rehabilitating stray dogs and cats. Volunteering at these organisations offers a chance to make a tangible difference in the lives of animals while connecting with other animal lovers. Other notable organisations include **BEES Elephant Sanctuary** and **Care for Dogs** in Chiang Mai, as well as the **Gibbon Rehabilitation Project** in Phuket.
- **Beach cleanups:** Thailand's beautiful beaches are a major attraction, but they also need protection from pollution. Organisations like **Trash Hero Thailand** and **Phuket Cleanse** organise regular beach cleanups. Participating in these events helps keep Thailand's beaches clean and safe while connecting you with other environmentally conscious individuals.
- **Community development:** Various NGOs and community organisations in Thailand work on projects to improve the lives of underprivileged communities. **Friends for Asia** and **GVI** are two

organisations that offer volunteer programmes focused on community development. Activities may include building infrastructure, organising community events, and supporting local families. Volunteering in these projects allows you to work closely with local communities and gain a deeper understanding of their challenges and aspirations. In Bangkok, **Operation Smile Thailand** also offers community-focused volunteer opportunities.

- **Education and teaching English:** Many schools and community centres in Thailand welcome volunteers to teach English and other subjects. Organisations like **Volunteers in Asia (VIA)** and **Projects Abroad** offer structured programmes where you can help improve students' language skills and broaden their horizons. Teaching English not only benefits the local community but also provides a unique opportunity to engage with Thai culture and learn more about the country's educational system. In Bangkok, organisations like **Scholars of Sustenance, The Mercy Centre, Friends for Asia**, and **Operation Smile Thailand** also offer education-focused volunteer programmes.
- **Environmental conservation:** Thailand's natural beauty and biodiversity make environmental conservation a significant focus for many organisations. **Trash Hero Thailand** is a grassroots movement that organises regular beach cleanups and environmental awareness campaigns. Volunteering with Trash Hero or similar organisations lets you contribute to preserving Thailand's pristine landscapes while meeting like-minded individuals who share your passion for the environment. The **Phuket Sunshine Village Foundation** also offers opportunities to get involved in environmental conservation efforts.

- **Sports coaching:** If you have a background in sports, consider volunteering as a sports coach. Many schools and community centres in Thailand welcome volunteers to help coach various sports, from football to tennis. This is a great way to stay active while making a positive impact on local youth.

**Local clubs**

Local clubs and organisations provide excellent opportunities to pursue hobbies, stay active, and meet new people. Whether you're interested in sports, arts, or cultural activities, there's likely a club in Thailand that aligns with your interests.

- **Arts and cultural clubs:** If you are passionate about the arts, joining a local arts or cultural club can be a rewarding experience. **Bangkok Community Theatre** is a popular group that welcomes expats and locals interested in theatre production and performance. Participating in their productions allows you to showcase your talents while making friends with fellow theatre enthusiasts. Similarly, **Chiang Mai Art Conversation** is an active community of artists and art lovers organising exhibitions, workshops, and cultural events. Joining such groups helps you explore Thailand's vibrant arts scene and connect with creative individuals.
- **Hiking and camping groups:** Thailand's diverse landscapes offer numerous opportunities for outdoor adventures. Joining hiking and camping groups can help you explore the country's natural beauty while meeting like-minded individuals. Groups like **Bangkok Hikers, Phuket Hiking Club,** and **Chiang Mai Hiking Club** organise regular trips to national parks, mountains, and other scenic locations. These excursions provide a

great way to stay active, enjoy nature, and build lasting friendships.

- **Language exchange groups:** Learning Thai can significantly enhance your experience in the country, and language exchange groups provide a fun and interactive way to improve your skills. Platforms like **Meetup.com** and **Facebook** host various language exchange events where you can practise Thai with native speakers and help them improve their English. These gatherings often occur in casual settings like cafés and bars, making them enjoyable social events.
- **Running and cycling clubs:** Thailand's scenic landscapes and well-maintained parks make it a great place for running and cycling enthusiasts. Clubs like **Bangkok Runners** and **Chiang Mai Cycling Club** organise regular group runs and rides, providing a supportive community for fitness enthusiasts. These clubs often participate in local races and events, offering additional opportunities to connect with fellow runners and cyclists.
- **Sailing Clubs:** Thailand's coastal waters and beautiful islands make it an excellent location for sailing enthusiasts. Joining a sailing club is a fantastic way to enjoy the marine environment, improve your sailing skills, and meet fellow sailors. Check out **Phuket Yacht Club** and **Royal Varuna Yacht Club Pattaya**.
- **Sports clubs:** Joining a sports club is a fantastic way to stay fit and socialise. Many cities in Thailand have clubs for various sports, including football, tennis, and Muay Thai. For example, the **Bangkok Lions Rugby Club and Samui Tennis Club** welcome locals and expats to join their clubs and participate in friendly matches and tournaments. If you're interested in martial arts, Muay Thai gyms nationwide offer classes for all skill levels.

Joining a Muay Thai gym not only helps you stay in shape but also provides an authentic cultural experience.

**Cultural Exchange Programmes**

Engaging in cultural exchange programmes is a fantastic way to immerse yourself in Thai culture, learn new skills, and build meaningful relationships with locals. These programmes often involve living with Thai families, participating in traditional activities, and contributing to community projects. Whether you're interested in learning Thai, exploring traditional arts, or experiencing rural life, there are numerous cultural exchange programmes that offer unique and enriching experiences. Let's dive into some of the best options for cultural exchange in Thailand.

**Homestays**

Homestay programmes provide an immersive experience where you live with a Thai family and participate in their daily activities. This is an excellent way to learn about Thai customs, cuisine, and lifestyle firsthand.

- **Cultural Homestay International (CHI):** CHI offers various homestay programmes in Thailand, where you can live with a Thai family and engage in cultural activities. These programmes often include language lessons, cooking classes, and excursions to local attractions. The immersive nature of a homestay provides a deep understanding of Thai culture and fosters strong connections with your host family.
- **Global Homestay Network (GHN):** This network offers homestay experiences with local families throughout Thailand. Participants can immerse themselves in Thai culture through daily activities, family meals, and cultural outings. The programme aims

to create lasting bonds and provide a genuine experience of Thai life.

- **XploreAsia Thailand Homestay Program:** This program offers travellers the unique opportunity to live with Thai families in various regions of Thailand. Participants can experience the local way of life, including traditional customs, foods, and daily activities. The program includes cultural orientations and support to ensure a fulfilling and enriching stay.

## Language immersion

Language immersion programmes offer structured lessons combined with cultural activities to help you learn the language and understand Thai culture.

**Bangkok**

**AAA Thai Language School**

- Based in Bangkok, this school provides a comprehensive immersion program that includes language lessons paired with cultural activities such as Thai cooking classes, traditional dance, and excursions to famous temples.

**AUA Language Center**

- The AUA Language Center in Bangkok offers immersive Thai language courses that focus on practical communication skills. The centre also organises cultural activities, such as cooking classes and temple visits, to complement your language learning. Participating in these activities allows you to practise your language skills in real-life situations and interact with locals
- **Duke Language School**
- Located in the heart of Bangkok, Duke Language School offers a variety of Thai language courses, including

immersive programs that blend classroom learning with cultural activities like cooking classes and day trips to local markets and temples.

- **Speak Thai Easy**
- An intimate language school in Bangkok that focuses on small class sizes and personalised attention. They offer immersive experiences through field trips to local markets, museums, and cultural landmarks.
- **Sumaa Language and Culture Institute**
- Situated in Bangkok, Sumaa offers an immersive Thai language program that combines structured lessons with cultural experiences like cooking classes, traditional dance lessons, and visits to historical sites.
- **Thai Language Hut**
- This Bangkok-based language school offers intensive Thai language courses that incorporate cultural activities to enhance learning. Students can participate in cooking classes, market visits, and cultural excursions to practice their language skills in authentic settings.

**Chiang Mai**

- **AUA Chiang Mai**
- Known for its well-structured courses and cultural immersion programs, AUA Chiang Mai offers practical Thai language learning combined with activities like market visits and traditional craft classes.
- **Chiang Mai University Language Institute**
- Located at Chiang Mai University, this institute offers comprehensive Thai language courses tailored to all proficiency levels. Alongside language classes, students can participate in cultural activities such as traditional dance workshops, local market tours, and cooking classes to enhance their immersion experience.
- **ProLanguage School**
- Located in Chiang Mai, ProLanguage provides a variety of Thai language courses with immersive cultural activities.

**Hua Hin**

**Sirada – The Learning Center**

- The school provides a dynamic learning environment for all levels, with additional cultural activities such as Thai cooking classes, local excursions, and traditional dance workshops to enhance the cultural experience.
- **The Language Centre Hua Hin**
- Conveniently provides various Thai language courses in the heart of the city. With an option to engage in cultural activities such as market visits and local tours.

**Pattaya**

- **AAA Pattaya School**
- Like its Bangkok sister, this school offers comprehensive language courses and cultural activities such as traditional dance lessons and excursions.
- **PLC Language School Pattaya**
- This school provides Thai language programs for various proficiency levels, integrating cultural activities such as cooking classes and local tours to enhance the learning experience.
- **ProLanguage School**
- This school offers a wide range of Thai language courses and cultural immersion activities at its Pattaya location.

**Phuket**

- **Alpha Language School**
  Recommended for its commitment to helping students achieve fluency in Thai. While they offer courses for beginners, their intensive programs are particularly good for helping learners make rapid progress.
- **Oriental Bamboo Language School**
  Known for its engaging Thai language programs, this school includes cultural immersion activities such as temple visits, traditional Thai crafts workshops, and local market tours.

- **Patong Language School**
- Offers comprehensive Thai language courses with cultural activities to enhance learning, such as cooking classes and city tours.

**Samui**

- **Mind Your Language**
- Provides tailored Thai language programs and cultural immersion activities, including visits to local markets and temples.
- **Samui Language School**
- Specialises in practical Thai language courses supplemented with cultural immersion activities like island tours and cooking classes.

**Online**

- **Learn Thai from a White Guy**
- Offers flexible online courses that combine language lessons with insights into Thai culture.
- **PickupThai Podcast**
- Offers a variety of immersive online Thai courses focusing on different aspects of conversational Thai and Thai culture.
- **Thaipod101**
- Provides extensive online courses with interactive lessons and cultural insights to help learners understand the language in context.

**Traditional arts and crafts**

Thailand has a rich heritage of traditional arts and crafts, and participating in these programmes allows you to learn new skills while appreciating the country's cultural heritage.

- **Baan Kang Wat**
- Located in Chiang Mai, Baan Kang Wat is an artist village offering various workshops and classes in traditional Thai crafts. From pottery and weaving to printmaking, this

community-driven space allows participants to engage with local artisans, learn new skills, and appreciate the depth of Thai craftsmanship. Visitors can also explore the creative boutique shops and quaint coffee shops within the village, making for a full artistic experience.

- **Baan Noorg Collaborative Arts and Culture**
- This organisation, based in Ratchaburi, offers workshops and residencies focusing on traditional and contemporary Thai arts. Participants can learn techniques in painting, sculpture, and performance art, fostering a deeper connection with Thai cultural expressions. The collaborative environment encourages creativity and exchange among artists, providing a unique opportunity to immerse yourself in the local art scene.
- **The Jim Thompson Art Center**
- Situated in Bangkok, The Jim Thompson Art Center offers a range of workshops and exhibitions focused on traditional Thai textiles and silk weaving. Participants can learn about the history and techniques of Thai silk production, as well as engage in hands-on weaving workshops. The art centre is part of the Jim Thompson House, a museum dedicated to the American entrepreneur who revitalised the Thai silk industry, providing a rich context for the learning experience.

### Community-based tourism

Community-based tourism (CBT) programmes focus on sustainable and responsible travel experiences that benefit local communities. These programmes often involve staying in rural villages, participating in community projects, and learning about local customs.

- **Baan Talae Nok Village:** Located on the Andaman Coast, Baan Talae Nok offers CBT experiences where visitors can stay with local families and engage in activities such as traditional fishing, batik painting, and mangrove restoration. The programme promotes sustainable tourism and supports the local community.

- **Koh Yao Noi Community-Based Tourism Club:** This club offers CBT experiences on the island of Koh Yao Noi, between Phuket and Krabi. Visitors can stay with local families, participate in fishing and farming activities, and learn about traditional boat building. The programme includes eco-tours, such as kayaking through mangroves and exploring coral reefs. Engaging in these activities helps preserve the island's natural environment and supports sustainable tourism.
- **Mae Kampong Village:** Located in the mountains of Chiang Mai, Mae Kampong Village offers a unique CBT experience. Visitors can stay in traditional teak houses, participate in tea and coffee production, and learn about local conservation efforts. The programme includes cultural activities, such as traditional dance performances and handicraft workshops. Staying in Mae Kampong Village provides an authentic glimpse into rural Thai life and supports the local community.

**Student exchange**

For younger digital nomads or those travelling with families, student exchange programmes offer a valuable opportunity for cultural immersion and academic growth.

- **AFS Intercultural Programs:** AFS offers student exchange programmes that place students with host families in Thailand. Participants attend local schools and engage in cultural activities, providing a comprehensive cultural exchange experience. The programme fosters cross-cultural understanding and builds lifelong connections.

- **Rotary Youth Exchange:** Rotary's Youth Exchange programme offers high school students the chance to spend a year living with host families in Thailand. Participants attend school, learn Thai, and take part in community service projects. The programme promotes cultural exchange and personal development, making it an enriching experience for young travellers.
- **Youth For Understanding (YFU):** YFU offers student exchange programmes that immerse participants in Thai culture by placing them with host families and enrolling them in local schools. The programme includes cultural activities and excursions, fostering a deep understanding of Thai life.

**Conclusion**

Joining local communities through volunteering, clubs, and cultural exchange programmes can significantly enhance your experience in Thailand. From teaching English and participating in sports clubs to living with Thai families and learning traditional crafts, there are countless ways to immerse yourself in Thai culture and build meaningful relationships. By actively engaging with the local community, you'll gain a deeper understanding of Thailand's rich heritage, contribute to the well-being of local communities, and create lasting memories. So, dive into these opportunities, embrace the cultural exchange, and make the most of your time in the Land of Smiles!

## SOCIAL EVENTS

Thailand isn't just a paradise for digital nomads because of its beautiful landscapes and affordable living; it's also a hotspot for social events and gatherings. Engaging in local social events is an excellent way to immerse yourself in Thai culture, meet new people, and build lasting memories. From vibrant festivals and traditional celebrations to modern meetups and networking events, there's always something happening in the Land of Smiles. In this section, we'll explore must-attend events, festivals, and meetups you should mark on your calendar.

## Must-Attend Events and Festivals

Thailand's social calendar is packed with a variety of events that cater to different interests and preferences. Whether you want to experience traditional Thai culture or connect with other sporting enthusiasts, there's an event for you. Here are some must-attend events and festivals you won't want to miss.

### Traditional festivals

Thailand is known for its colourful and lively festivals, which offer a unique glimpse into the country's rich cultural heritage. Participating in these festivals is a fantastic way to experience Thai traditions and celebrate with locals. For more, refer to **Chapter 6: Cultural Insights & Learning Thai.**

### Food and drink festivals

Thailand's world-renowned culinary scene and food and drink festivals are a fantastic way to sample the country's diverse flavours. These events feature local and international cuisine, cooking demonstrations, and live entertainment.

- **Bangkok Street Food Festival:** This festival annually celebrates Bangkok's vibrant street food culture. Vendors from across the city gather to showcase their best dishes, offering everything from spicy papaya salad to sweet mango sticky rice. The festival also features cooking competitions and cultural performances, making it a fun and delicious experience.
- **Chiang Mai Street Food Festival:** Similar to its Bangkok counterpart, the Chiang Mai Street Food Festival highlights the region's unique culinary

traditions. The event brings together street food vendors, chefs, and enthusiasts to celebrate Northern Thai cuisine. Attendees can sample various dishes, watch cooking demonstrations, and enjoy live music and entertainment.

- **Hua Hin Food Festival:** The Hua Hin Food Festival is a popular event that showcases the best of the region's culinary offerings. Visitors can enjoy a variety of dishes from local vendors, participate in cooking classes, and watch live cooking demonstrations. The festival also features cultural performances and entertainment, making it a great family-friendly event.
- **Pattaya International Food and Hot Air Balloon Festival:** This festival combines culinary delights with the spectacle of hot air balloons. Held annually in Pattaya, it features a wide array of food stalls offering Thai and international cuisine. The event also includes cooking demonstrations, live music, and the opportunity to ride in a hot air balloon.
- **Phuket Seafood Festival:** Phuket is an island destination that is famous for its fresh seafood. The Phuket Seafood Festival showcases the best of the island's seafood offerings, with local restaurants and vendors serving up delicious dishes. The festival also includes cooking demonstrations, seafood-themed competitions, and cultural performances, making it a must-visit for seafood lovers.

**Fitness events**

For those interested in health and wellness, Thailand offers a range of events focused on fitness, mindfulness, and holistic living. These events provide opportunities to improve your well-being, learn new practices, and connect with like-minded individuals.

- **Bangkok Marathon:** For fitness enthusiasts, the Bangkok Marathon is a major event that attracts runners from around the world. The marathon offers various race categories, including a full marathon, half marathon, and fun run. Participating in the Bangkok Marathon is a great way to challenge yourself, stay active, and experience the city's vibrant energy.
- **Ironman Thailand:** The Ironman Thailand event, typically held in Phuket, is part of the global Ironman series. This triathlon includes a 2.4-mile swim, a 112-mile bike ride, and a 26.2-mile marathon. It attracts athletes worldwide who are looking to push their limits and achieve their fitness goals. The event is known for its scenic routes and supportive atmosphere.
- **Phuket Triathlon:** The Laguna Phuket Triathlon is one of Asia's premier triathlon (swim, bike, run) events, featuring a 1.8-kilometre swim, a 50-kilometre bike ride, and a 12-kilometre run. The event is renowned for its beautiful course, which takes participants through tropical landscapes and along pristine beaches. It's an excellent opportunity for triathletes to compete in a world-class event.
- **Spartan Race:** Held in various locations across Thailand, the Spartan Race is an obstacle course race that tests your endurance, strength, and determination. Participants can choose from different race categories, including the Sprint, Super, and Beast courses. The event is a fantastic way to challenge yourself physically and mentally while connecting with other fitness enthusiasts.
- **Thailand Yoga Festival:** Held annually in various locations, the Thailand Yoga Festival brings together yoga enthusiasts, instructors, and wellness practitioners for a weekend of yoga classes, workshops, and holistic

activities. The festival offers a diverse range of yoga styles, meditation sessions, and wellness talks, making it an enriching experience for both beginners and advanced practitioners.

**Cultural and arts events**

Thailand's rich cultural heritage and thriving arts scene are showcased through various events and festivals celebrating music, dance, theatre, and visual arts.

- **Bangkok Art Biennale:** The Bangkok Art Biennale is an international contemporary art festival that takes place every two years. The event features works by artists worldwide, displayed in various venues across the city, including temples, galleries, and public spaces. The Biennale offers a unique opportunity to explore contemporary art and engage with the local and international art community.
- **Bangkok International Festival of Dance & Music:** This annual festival features performances by renowned international and local artists, showcasing various genres, including ballet, opera, classical music, and contemporary dance. The event is held at the Thailand Cultural Centre and attracts global art enthusiasts.
- **Chiang Mai Flower Festival:** Held in February, the Chiang Mai Flower Festival is a stunning celebration of the region's floral beauty. The festival features elaborate flower displays, parades with flower-covered floats, and cultural performances. Attending the Chiang Mai Flower Festival provides a colourful and fragrant experience highlighting the city's horticultural heritage.
- **Pattaya International Music Festival:** This annual festival in Pattaya attracts local and international musicians, offering a diverse lineup of performances across various genres. The festival occurs on multiple

stages along the beach, creating a lively and festive atmosphere. Whether you're a fan of rock, pop, jazz, or electronic music, the Pattaya International Music Festival has something for everyone.

- **Wonderfruit:** This annual festival, held in Pattaya, combines music, art, and sustainable living. Wonderfruit features performances by international and local musicians, art installations, workshops, and wellness activities. The festival promotes environmental consciousness and offers a unique, immersive experience for attendees.

**Technology and innovation events**

Thailand is rapidly becoming a hub for technology and innovation, hosting an array of events that bring together entrepreneurs, investors, and tech enthusiasts. These events provide opportunities to learn about the latest advancements, network with industry leaders, and explore new business opportunities.

- **ASEAN Angel Alliance Summit (AAAS):** Held in Bangkok, the AAAS brings together angel investors, startup founders, and ecosystem builders across the ASEAN region. The event focuses on fostering investment and collaboration, offering pitch sessions, panel discussions, and networking opportunities.
- **Echelon Thailand:** Echelon is a regional tech conference that takes place in various Southeast Asian countries, including Thailand. The event focuses on technology, startups, and entrepreneurship, featuring keynote speeches, panel discussions, and networking opportunities.
- **Startup Thailand:** This annual event, organised by the Thai government, aims to support and promote the startup ecosystem in Thailand. The event includes

exhibitions, pitch competitions, workshops, and networking opportunities, attracting startups, investors, and industry experts worldwide.

- **Techsauce Global Summit:** Held annually in Bangkok, the Techsauce Global Summit is one of Asia's leading technology conferences. It gathers entrepreneurs, investors, and professionals from various industries to discuss the latest trends in technology and innovation. The summit features keynote speakers, panel discussions, and networking sessions.
- **Thailand Digital Big Bang:** Organised by the Ministry of Digital Economy and Society, this event showcases Thailand's advancements in digital technology. The event includes exhibitions, seminars, and workshops on AI, IoT, and cybersecurity, providing valuable insights and networking opportunities for tech enthusiasts.

## Making the Most of Social Events

Attending social events in Thailand can greatly enhance your experience, but planning ahead is key to making the most of them. Here are some practical tips to ensure you don't miss out on the best events and activities.

### Being prepared

**Buy tickets in advance:** Popular events and festivals in Thailand can sell out quickly, especially during peak tourist seasons. To secure your spot, buy tickets in advance whenever possible. Websites like **Ticketmelon, Eventpop,** and **ThaiTicketMajor** offer online ticketing services for a wide range of events.

**Check event dates:** Before making travel plans or scheduling other activities, check the dates of upcoming events. Websites like **Eventbrite**, **Meetup**, and **Facebook Events** are excellent resources for finding event dates and details. Additionally, local tourism websites and event calendars can provide valuable information.

**Join groups and communities:** Joining online groups and communities can help you stay informed about upcoming events and activities. Platforms like **Facebook**, **Meetup**, and **LINE** have numerous groups dedicated to various interests, from fitness and food to technology and arts. Joining these groups allows you to receive event updates, participate in discussions, and connect with other attendees.

**Set notifications:** To ensure you don't miss out on important events, set notifications on your phone or computer. Many event websites and apps offer notification features that alert you when tickets go on sale or when event dates are approaching. Additionally, subscribing to newsletters from event organisers can keep you informed about upcoming activities.

**Plan your schedule:** Once you have a list of events you want to attend, plan your schedule to make the most of your time. Consider the locations and timings of events to avoid overlaps and ensure you have enough time to enjoy each activity. Using a digital calendar or planner can help you stay organised and manage your time effectively.

### Networking 101

Networking is an essential aspect of thriving as a digital nomad. Building a strong network can lead to new opportunities, collaborations, and lasting friendships. Here are some tips for effective networking:

- **Attend networking events:** Look for events specifically designed for networking, such as business mixers, industry conferences, and professional meetups. These events provide a platform to meet potential collaborators, clients, and mentors.

- **Follow up:** After meeting new contacts, make sure to follow up with a message or email. This helps solidify the connection and opens the door for future interactions.
- **Join professional groups:** Many cities have professional groups and associations based on various industries. Joining these groups can provide ongoing networking opportunities and access to valuable resources.
- **Leverage LinkedIn:** Use LinkedIn to connect with people you meet at events. It's a professional platform that allows you to maintain and grow your network.
- **Prepare your pitch:** Have a clear and concise pitch ready about who you are, what you do, and what you're looking for. This helps make a strong first impression and facilitates meaningful conversations.

By incorporating these strategies, you can maximise your social and professional experiences in Thailand, ensuring that you make the most of your time in the Land of Smiles.

## Conclusion

Thailand's social events offer a wealth of opportunities to immerse yourself in the local culture, meet new people, and create unforgettable memories. From traditional festivals and bustling food markets to fitness events and cultural performances, there's always something happening in the Land of Smiles. By planning ahead, joining groups, and setting notifications, you can ensure you don't miss out on any exciting activities. Embrace the vibrant social scene, connect with like-minded individuals, and make the most of your digital nomad journey in Thailand. With a bit of preparation and enthusiasm, you'll find that Thailand offers endless possibilities for enriching your social life and making lasting connections.

# MAINTAINING CONNECTIONS

One of the most significant challenges digital nomads face is maintaining connections with family and friends back home. While the allure of exploring new places and immersing yourself in new cultures is exhilarating, staying connected to your roots is essential for emotional support and overall well-being. Fortunately, technology has made it easier than ever to keep in touch, no matter where you are in the world. In this section, we'll explore various strategies and tools for staying connected with loved ones back home, ensuring that your relationships remain strong and supportive even as you embark on your digital nomad journey.

## Staying in Touch with Loved Ones Back Home

Maintaining regular communication with family and friends can help alleviate feelings of loneliness and homesickness. Here are some effective ways to stay connected and ensure that your relationships remain robust, even when you are miles away.

### Video calls and messaging apps

Video calls and messaging apps are the lifelines for digital nomads. These tools allow real-time communication, making it feel like you're just a click away from your loved ones.

- **FaceTime:** If you and your family use Apple devices, FaceTime is a seamless option for video calls. Its integration with iOS makes it easy to initiate calls directly from your contacts list. FaceTime supports both one-on-one and group calls.
- **Google Duo:** Another reliable option for video calls is Google Duo. It's available on both Android and iOS devices and offers high-quality video and audio. Google Duo also supports group calls, making it a convenient choice for connecting with multiple people at once.

- **LINE:** This popular Asian app is about more than just messaging. LINE offers text messaging, voice and video calls, and various social networking features such as timelines, stickers, and games. It's widely used in Thailand and is great for staying connected with locals and expats alike.
- **Skype:** Skype has been a go-to for international communication for years. It offers video calls, voice calls, and instant messaging. Skype also provides the option to call landlines and mobile numbers at affordable rates, making it a versatile tool for staying in touch.
- **WhatsApp:** This popular messaging app offers text messaging and video calling features. It's widely used around the globe, making it an excellent choice for keeping in touch with family and friends. WhatsApp also supports group chats, allowing you to stay connected with multiple people simultaneously.
- **Zoom:** Known primarily for its use in professional settings, Zoom is also an excellent tool for personal video calls. Its high-quality video and audio make it ideal for catching up with loved ones. Zoom offers both free and paid plans, with the free plan allowing unlimited one-on-one calls and 40-minute group calls.

**Social media platforms**

Social media platforms are not only great for staying updated with what's happening back home but also for sharing your own experiences and adventures. Here are some popular platforms to consider.

- **Facebook:** Facebook remains one of the most widely used social media platforms. It allows you to share updates, photos, and videos with your network. Facebook's Messenger app also supports text messaging and video calls, providing multiple ways to stay in touch.

- **Instagram:** Instagram is perfect for sharing visual content. You can post photos and videos of your travels, use Instagram Stories for real-time updates, and even go live to interact with your followers. Direct messaging on Instagram also allows you to have private conversations with friends and family.
- **LinkedIn:** For professional connections, LinkedIn is the go-to platform. While it's primarily used for networking and career-related updates, it also allows you to share personal milestones and achievements with your professional network.
- **Twitter:** If you prefer short and concise updates, Twitter is an excellent platform. You can tweet about your experiences, share photos, and engage with your followers through replies and direct messages.

**Email communication**

While instant messaging and social media are great for quick updates, email remains an effective way to share more detailed and thoughtful messages. Here are some tips for using email to stay connected.

- **Interactive content:** Include interactive content like questionnaires, polls, or simple surveys within your emails. This can make your communication more engaging and encourage your family and friends to interact with you more actively, creating a deeper sense of connection.
- **Personalised messages:** Take the time to write personalised messages to your close family members and friends. This shows that you value your relationship and are trying to stay connected despite the distance.

- **Photo attachments:** Attach photos to your emails to provide a visual element to your updates. This helps your loved ones feel more connected to your experiences and adventures.
- **Regular updates:** Consider sending regular email updates to your family and friends. These could be monthly newsletters or personal messages that detail your experiences, challenges, and achievements. This format allows you to share more in-depth reflections than social media posts.

### Care packages and snail mail

While digital communication is convenient, there's something special about receiving a physical letter or package. Sending and receiving care packages and handwritten letters can add a personal touch to your long-distance relationships.

- **Care packages:** Consider sending care packages with local souvenirs, snacks, and other small items that reflect your travel experiences. Receiving a care package can brighten someone's day and provide a tangible connection to your journey.
- **Handwritten letters:** Taking the time to write and send a handwritten letter can profoundly impact. It's a thoughtful gesture that shows you care and are willing to put in the effort to stay connected.
- **Personalised gifts:** Tailor your gifts to the interests and preferences of the recipient. Whether it's a book from your travels, a piece of local art, or a handcrafted item, personalised gifts demonstrate thoughtfulness and show that you truly understand and care about the person on the receiving end.
- **Postcards:** Sending postcards from your travel destinations is a simple yet meaningful way to stay

connected. It’s a tangible reminder of your adventures and shows that you’re thinking of your loved ones.

**Time management and scheduling**

Maintaining regular communication with family and friends requires effective time management and scheduling. Here are some tips to help you stay on track.

- **Account for time zones:** Be mindful of time zone differences when scheduling calls and sending messages. Use tools like World Time Buddy to find suitable times for both parties.
- **Balance work and personal time:** Make sure to allocate time for personal communication in your schedule. Balancing work and personal time ensures you stay connected without compromising your productivity.
- **Set regular call times:** Establish a routine for video calls and stick to it. This could be a weekly or bi-weekly call at a time that works for both you and your loved ones. Having a set schedule helps ensure that you stay connected consistently.
- **Use reminders and calendar apps:** Utilise digital tools such as reminders and calendar apps to set alerts for scheduled communications. This can help you stay on top of your commitments and ensure you don't miss important calls or messages, thus maintaining a consistent connection with your loved ones.

**Shared experiences and activities**

Sharing experiences and activities with your loved ones can help bridge the distance and strengthen your connections.

- **Book clubs:** Start a virtual book club with your family and friends. Choose a book to read each month and meet online to discuss it. This creates a shared intellectual activity and provides ample material for engaging conversations, deepening your connection over a mutual interest.

- **Online games:** Playing online games together can be a fun and interactive way to connect. Whether it's a casual game of Words With Friends or a more involved multiplayer game, gaming provides an opportunity for shared enjoyment.
- **Virtual cooking:** Arrange virtual cooking sessions where you and your loved ones prepare the same recipe together over a video call. This can be a fun and engaging way to share a meal and catch up on each other's lives.
- **Watch parties:** Use platforms like Netflix Party or Amazon Prime Watch Party to watch movies and TV shows together in real-time. This allows you to share the experience and discuss the content as you watch.

## Making Long-Term Friendships

While staying in touch with family and friends back home is essential, building new, long-term friendships in your new location can significantly enhance your experience as a digital nomad. Long-term friendships provide community, support, and belonging, making your time in Thailand more enjoyable and fulfilling. In this section, we'll explore strategies for making long-term friendships, from engaging in local activities to nurturing relationships over time.

### Engaging in local activities

One of the best ways to meet new people and form long-term friendships is by actively participating in local activities and events. Here are some suggestions to help you get started.

- **Attend meetups and social events:** Platforms like **Meetup.com** and **Facebook Events** are great resources for finding social gatherings and meetups in your area.

These events often cater to various interests, from professional networking to casual socialising. Attending meetups allows you to expand your social circle and meet potential long-term friends.

- **Join clubs and organisations:** Whether you're interested in sports, arts, or hobbies, joining local clubs and organisations is an excellent way to meet like-minded individuals. For example, if you enjoy running, consider joining a local running club like **Bangkok Runners** and **Hash House Harriers**. If you're passionate about the arts, look for groups like **Creative Chiang Mai**. If you want to improve your social and public speaking skills, join **InterNations** or **Toastmasters International**. Participating in these activities regularly helps you build connections with others who share your interests.
- **Take local classes or workshops:** Sign up for classes or workshops in your area that cater to your interests, such as cooking, painting, or language courses. These educational settings provide a structured environment to learn something new while meeting people with similar interests. Engaging in a shared activity facilitates easier conversation and helps foster new friendships.
- **Volunteer:** Volunteering for local organisations and charities not only allows you to give back to the community but also provides opportunities to meet people who share your values. Look for volunteering opportunities in areas that interest you, such as animal welfare, education, or environmental conservation. Volunteering regularly helps you form deeper connections with fellow volunteers.

### Building meaningful connections

Forming long-term friendships requires more than just meeting new people; it involves building meaningful connections based on trust, mutual interests, and shared experiences. Here are some tips to help you cultivate these connections.

- **Be open and approachable:** Approach new social situations with an open mind and a friendly attitude. Smile, make eye contact, and engage in conversations with genuine interest. Being approachable makes it easier for others to connect with you and helps create a positive impression.
- **Listen actively:** Active listening is a crucial skill for building meaningful connections. Respond to what others say, ask follow-up questions, and show empathy. Demonstrating that you care about their thoughts and feelings helps build trust and rapport.
- **Share your experiences:** Don't be afraid to share your experiences and stories. Being open about your journey, challenges, and achievements allows others to relate to you and fosters a sense of camaraderie. Sharing personal experiences can also lead to deeper, more meaningful conversations.
- **Participate in group activities:** Group activities provide opportunities for shared experiences, essential for building long-term friendships. Whether it's joining a group hike, attending a cooking class, or participating in a language exchange, engaging in group activities helps you bond with others over common interests.

### Nurturing friendships over time

Building long-term friendships requires ongoing effort and nurturing. Here are some strategies to help you maintain and strengthen your friendships over time.

- **Stay in touch:** Regular communication is key to maintaining friendships. Use messaging apps, social media, and video calls to stay connected with your friends. Even a simple message or a quick call can go a long way in showing that you care and are thinking of them.
- **Plan regular meetups:** Make an effort to plan regular meetups with your friends, whether it's a weekly coffee catch-up, a monthly dinner, or a weekend outing. Consistent meetups help reinforce your bond and create opportunities for shared experiences.
- **Celebrate milestones:** Acknowledge and celebrate important milestones in your friends' lives, such as birthdays, anniversaries, or achievements. Small gestures like sending a birthday card or organising a surprise celebration show that you value and appreciate your friendship.
- **Be supportive:** Offer support and encouragement during good times and challenging moments. Being there for your friends when they need you helps build trust and strengthens your bond. Whether it's offering a listening ear, providing practical assistance, or simply being present, your support can make a significant difference.
- **Show appreciation:** Express gratitude and appreciation for your friends and the positive impact they have on your life. Simple acts of kindness, such as saying thank you, giving compliments, or doing something thoughtful, can help reinforce your connection and make your friends feel valued.

**Connecting with locals and expats**

Building long-term friendships with locals and fellow expats can enrich your experience and provide a well-rounded support network. Here's how to connect with both groups.

- **Engage with locals:** Making local friends allows you to gain a deeper understanding of Thai culture and traditions. Attend cultural events, learn the language, and show genuine interest in the local way of life. Engaging with locals helps bridge cultural gaps and fosters meaningful connections.
- **Join expat communities:** Expat communities provide a valuable support network for navigating life in a new country. Joining expat groups, attending events, and participating in online forums can help you connect with others with similar experiences and challenges. Platforms like **InterNations** and local **Facebook Groups** for expats are excellent resources for finding these communities.
- **Participate in language exchange programmes:** Language exchange programmes are a great way to meet both locals and expats while improving your language skills. These programmes often involve casual meetups where participants practise their target language with native speakers. This setting provides a relaxed environment for making new friends.
- **Get involved in local sports and hobby clubs:** Joining local sports teams or hobby clubs is another excellent way to meet people. Whether it's a local football team, a yoga class, or a photography club, participating in these activities allows you to connect with others who share your interests and passions.

**Maintaining friendships when moving**

As a digital nomad, you may move frequently, but that doesn't mean you have to leave your friendships behind. Here are some tips for maintaining friendships even when you're on the move.

- **Stay connected online:** Use messaging apps, social media, and video calls to keep in touch with your friends, regardless of where you are. Regular communication helps maintain your bond and ensures that your friendship remains strong.

- **Plan visits:** If possible, plan visits to see your friends in person. Whether it's a weekend trip or a longer stay, spending time together in person helps reinforce your connection and create new memories.
- **Send updates:** Keep your friends updated on your travels and experiences. Sharing your journey through blog posts, social media updates, or personal messages helps them feel connected to your adventures.
- **Make new friends:** Embrace the opportunity to make new friends in each location you visit. Building a global network of friends enriches your experience and provides a diverse support system.

**Conclusion**

Maintaining connections as a digital nomad involves staying in touch with family and friends back home and building long-term friendships in your new location. By leveraging technology, engaging in local activities, and nurturing relationships over time, you can create a strong support network that enhances your overall well-being and enriches your experience in Thailand. Balancing old and new connections ensures that you stay emotionally grounded and enjoy a fulfilling social life, no matter where your digital nomad journey takes you. So, embrace the opportunities to connect, build lasting friendships, and make the most of your time in the Land of Smiles!

**Stay tuned . . .**

As we wrap up our journey through the vibrant and enriching experience of being a digital nomad in Thailand, there's one more crucial aspect to explore: family and schooling. In

**Chapter 10: Family & Schooling**, we'll dive into the world of education on the move, offering a comprehensive guide for nomad families. Discover the top international schools in Thailand, explore various curriculum options and enrolment processes, and weigh the pros and cons of local Thai schools. We'll also delve into homeschooling resources, providing tips on balancing travel and education, and highlight extra-curricular activities to keep your children engaged and socialised.

Whether you're travelling with young kids or teenagers, this chapter will equip you with the knowledge and resources to ensure your family thrives while you embrace the digital nomad lifestyle. Stay tuned for valuable insights and practical advice to make your family's educational journey as seamless and enriching as possible!

# Chapter 10:
# Family & Schooling

## Education on the Move: Schooling Options for Nomad Families

Embarking on a digital nomad journey with your family can be an incredibly rewarding experience, but it also comes with its unique set of challenges, especially regarding your children's education. Thailand offers a plethora of schooling options to ensure your kids receive a quality education while you explore the world. In this chapter, we'll guide you through the top international schools, delve into local Thai school alternatives, and explore the ins and outs of homeschooling on the move. We'll also highlight extra-curricular activities to keep your children engaged and socialised. Whether navigating enrolment processes, balancing travel and education, or looking for ways to integrate culturally, this chapter is packed with valuable insights and practical advice to help your family thrive. Get ready to discover how to turn your nomadic dream into a seamless educational adventure for your little ones!

### INTERNATIONAL SCHOOLS

When it comes to providing your children with a top-notch education while living as a digital nomad in Thailand, international schools are a fantastic option. These schools offer globally recognised curricula, excellent facilities, and a multicultural environment that can help ease the transition for your kids. International schools often have experienced teachers and a broad range of extra-curricular activities, ensuring a well-rounded education. Let's dive into some of the top international schools across Thailand that cater to the needs of expat families, offering both academic excellence and a supportive community for your children.

## Top International Schools in Thailand

### Bangkok

### Bangkok Patana School

Bangkok Patana School is one of the oldest and most prestigious international schools in Thailand. Established in 1957, it offers a British-based curriculum and caters to students from Nursery to Year 13. The school is known for its impressive facilities, which include state-of-the-art science labs, sports fields, and a performing arts centre. Bangkok Patana also places a strong emphasis on extra-curricular activities, providing a well-rounded education for its students.

### Harrow International School Bangkok

Harrow International School Bangkok is part of the prestigious Harrow family of schools, with a history dating back to 1572 in the UK. This school offers a British-based curriculum and caters to students from Early Years to Sixth Form. Harrow Bangkok is known for its rigorous academic standards, extensive extra-curricular programmes, and strong emphasis on leadership and character development. The school's sprawling campus includes world-class sports facilities, performing arts venues, and boarding options.

### International School Bangkok (ISB)

Located in the serene Nichada Thani community, International School Bangkok (ISB) is another top choice for expat families. ISB offers an American-based curriculum and the International Baccalaureate (IB) programme, catering to students from Pre-Kindergarten to Grade 12. The school boasts a diverse student body, with over 60 nationalities represented. ISB's campus is equipped with cutting-edge facilities, including a swimming pool, fitness centre, and a dedicated arts and music building.

**NIST International School**

NIST International School, situated in the heart of Bangkok, is renowned for its commitment to academic excellence and holistic development. NIST offers the full IB programme from the Primary Years Programme (PYP) to the Diploma Programme (DP). The school's facilities are top-notch, featuring modern classrooms, sports complexes, and innovative learning spaces. NIST also emphasises community service and sustainability, encouraging students to become responsible global citizens.

**Ruamrudee International School (RIS)**

Ruamrudee International School (RIS) is a well-established American curriculum school located in Minburi, Bangkok. Founded in 1957, RIS offers a comprehensive education from Pre-Kindergarten to Grade 12, incorporating Advanced Placement (AP) courses and the IB Diploma Programme. The school's diverse student body and experienced faculty create a nurturing environment for academic and personal growth. RIS's campus features modern facilities, including science labs, art studios, and sports amenities.

**Shrewsbury International School Bangkok**

Shrewsbury International School Bangkok is another top-tier institution offering a British-based curriculum. With two campuses – Riverside and City – Shrewsbury caters to students from Early Years to Sixth Form. The school is known for its strong academic performance, particularly in IGCSE and A-Level examinations. Shrewsbury's campuses are equipped with excellent facilities, including sports complexes, music studios, and science labs. The school also places a strong emphasis on extra-curricular activities, offering a wide range of clubs and sports teams.

## Chiang Mai

### Chiang Mai International School (CMIS)

For families based in the north of Thailand, Chiang Mai International School (CMIS) is an excellent option. CMIS offers an American-based curriculum and caters to students from Pre-Kindergarten to Grade 12. The school's diverse student body and experienced faculty create a supportive and inclusive learning environment. CMIS's campus includes state-of-the-art facilities, such as science labs, sports fields, and a performing arts centre. The school also offers many extra-curricular activities, including sports, arts, and community service programmes.

### Prem Tinsulanonda International School

Located just outside Chiang Mai, Prem Tinsulanonda International School is a highly regarded institution offering the IB programme from the Primary Years Programme (PYP) to the Diploma Programme (DP). The school's stunning campus is set amidst lush greenery and features world-class facilities, including sports complexes, art studios, and boarding options. Prem Tinsulanonda places a strong emphasis on holistic education, encouraging students to excel academically, socially, and personally.

### Panyaden International School

Panyaden International School is an eco-friendly institution located in Chiang Mai. It offers a unique curriculum that blends the British curriculum with Buddhist principles, catering to students from Nursery to Year 13. The school is known for its commitment to sustainability and mindfulness, providing a nurturing environment that focuses on both academic and personal growth. Panyaden's campus features natural building materials, open spaces, and modern learning facilities.

## Koh Samui

### Lanna International School Thailand

Lanna International School Thailand, located on the island of Koh Samui, offers a British-based curriculum and caters to students from Early Years to Sixth Form. The school is known for its strong academic standards, personalised learning approach, and vibrant community. Lanna's campus features modern facilities, including sports fields, science labs, and art studios.

### Oonrak International School

Oonrak International School, situated on the island of Koh Samui, offers a comprehensive education from Nursery to Year 13. The school provides a British-based curriculum and focuses on holistic development, ensuring that students excel academically, socially, and personally. Oonrak's campus includes state-of-the-art facilities, such as sports fields, swimming pools, and art studios.

### The Learning Tree International Kindergarten and School

The Learning Tree, located on the island of Koh Samui, offers an early year's education based on the British curriculum. The school caters to children from Nursery to Year 6 and is known for its nurturing environment and strong emphasis on play-based learning. The Learning Tree's campus features open play areas, modern classrooms, and dedicated learning spaces.

## Pattaya

### International School Eastern Seaboard (ISE)

Located in Chonburi, International School Eastern Seaboard (ISE) serves the expat community in the Eastern Seaboard region. ISE offers an American-based curriculum and the IB Diploma Programme, catering to students from Pre-Kindergarten to Grade 12. The school's small class sizes and personalised approach to education ensure that each student receives individual attention and support. ISE's campus features modern classrooms, sports facilities, and a dedicated arts centre.

### Regents International School Pattaya

Regents International School Pattaya offers a British-based curriculum and the IB Diploma Programme, catering to students from Early Years to Sixth Form. The school's campus includes state-of-the-art facilities such as sports fields, swimming pools, and art studios. Regents Pattaya is known for its strong academic performance and extensive extra-curricular programmes, which include sports, arts, and community service.

### St. Andrews International School, Green Valley

St. Andrews International School, Green Valley, is located in Rayong, near Pattaya. The school offers a British-based curriculum and caters to students from Nursery to Year 13. St. Andrews is known for its strong academic standards, personalised learning approach, and vibrant community. The school's campus features modern facilities, including sports fields, science labs, and art studios.

## Phuket

### British International School, Phuket (BISP)

British International School, Phuket (BISP) is another excellent option for families in the south. BISP offers a British-based curriculum and the IB Diploma Programme, catering to students from Early Years to Sixth Form. The school's impressive campus features state-of-the-art facilities, including sports complexes, swimming pools, and performing arts venues. BISP is known for its strong academic performance, particularly in

IGCSE and A-Level examinations, and its extensive extra-curricular programmes, including sports, arts, and community service.

### HeadStart International School

HeadStart International School, with two (City and Cherngtalay) campuses located in Phuket, is an excellent choice for families in the southern part of Thailand. The school offers the Cambridge International Curriculum, catering to students from Early Years to Sixth Form. HeadStart is known for its strong academic performance, particularly in IGCSE and A-Level examinations. The school's modern campuses are equipped with state-of-the-art facilities, including sports fields, swimming pools, and performing arts venues. HeadStart also places a strong emphasis on extra-curricular activities, offering a wide range of clubs and sports teams to keep students engaged and active.

### UWC Thailand International School

UWC Thailand International School, located in Phuket, is part of the prestigious United World Colleges (UWC) network. The school offers the IB programme from the Primary Years Programme (PYP) to the Diploma Programme (DP) and focuses on holistic education, sustainability, and global citizenship. UWC Thailand's campus includes modern facilities, such as sports fields, swimming pools, and art studios. The school's diverse student body and commitment to experiential learning create a dynamic and inclusive environment.

## Others

### Hua Hin International School

Hua Hin International School, located in Hua Hin, offers a British-based curriculum and the IB Diploma Programme, catering to students from Early Years to Sixth Form. The school is known for its strong academic standards, personalised learning approach, and vibrant community. Hua Hin's campus features modern facilities, including sports fields, science labs, and art studios.

### Krabi International School

Krabi International School, located in Krabi, offers a British-based curriculum and caters to students from Nursery to Year 13. The school is known for its commitment to academic excellence and holistic development. Krabi's campus includes state-of-the-art facilities, such as sports fields, swimming pools, and art studios.

### The International School of Asia

The International School of Asia, located in Koh Lanta, offers a British-based curriculum and caters to students from Nursery to Year 13. The school is known for its strong academic standards, personalised learning approach, and vibrant community. The school's campus features modern facilities, including sports fields, science labs, and art studios.

## Education Essentials

Choosing the right curriculum for your child is a crucial aspect of selecting an international school in Thailand. With various curricula available, such as British, American, and International Baccalaureate (IB), each school offers unique educational frameworks. Additionally, the enrolment processes and pricing structures can vary significantly. This chapter will break down the main curriculum options, guide you through the steps to enrol your child and provide insights into the costs involved. By understanding these factors, you can make an informed decision to ensure your child receives the best possible education while living in Thailand.

**Curriculum options**

International schools in Thailand typically offer a range of globally recognised curricula, each with its unique strengths and focus areas. Here are the most common options:

**British curriculum (IGCSE and A-Levels):** The British curriculum is widely respected and offers a structured approach to education. It includes the International General Certificate of Secondary Education (IGCSE) for students aged 14-16, followed by A-Levels for students aged 16-18. Schools like Bangkok Patana, Harrow International School, and Shrewsbury International School offer this curriculum, providing a solid foundation for university applications worldwide.

**American curriculum:** The American curriculum is another popular choice, focusing on various subjects and fostering critical thinking. It includes Advanced Placement (AP) courses, which allow students to earn college credits while in high school. International School Bangkok (ISB) and Ruamrudee International School (RIS) are known for their American-based programmes, offering a well-rounded education that prepares students for higher education in the US and beyond.

**International Baccalaureate (IB):** The IB programme is renowned for its rigorous academic standards and emphasis on holistic development. It includes the Primary Years Programme (PYP), Middle Years Programme (MYP), and Diploma Programme (DP). Schools like NIST International School, Prem Tinsulanonda International School, and UWC Thailand offer the full IB continuum, promoting critical thinking, intercultural understanding, and personal growth.

**Other curricula:** Some international schools in Thailand offer national curricula from other countries, such as the French Baccalaureate, German Abitur, or Australian Curriculum. These options may be suitable for families planning to return to their home country or seeking a specific educational approach. Additionally, alternative educational approaches, such as Montessori, Steiner, Finnish, and Waldorf, are also available in some schools. These methods focus on child-centred learning, creativity, and holistic development, providing different pathways for children's education.

**Enrolment process**

Enrolling your child in an international school in Thailand involves several steps, which can vary depending on the school. Here's a general overview of the enrolment process:

**Research and shortlist:** Start by researching schools that offer the curriculum and facilities you're looking for. Consider factors like location, extra-curricular activities, and community reviews. Once you've shortlisted a few schools, visit their websites to gather detailed information about their programmes and admission requirements.

**Application:** Most schools require you to complete an online application form, which typically includes basic information about your child, their academic history, and your family's background. You may also need to submit supporting documents, such as transcripts, recommendation letters, and a copy of your child's passport.

**Assessment and interview:** Many international schools conduct assessments and interviews as part of the enrolment process. These assessments may include academic tests, language proficiency evaluations, and interviews with the child and parents. The purpose is to gauge your child's readiness for the school's curriculum and ensure a good fit.

**Acceptance and enrolment:** If your child is accepted, you'll receive an offer letter with details about the enrolment process, including deadlines and required documents. You must pay an enrolment fee and submit additional paperwork, such as medical records and proof of residence.

**Orientation:** Once enrolled, your child will likely attend an orientation programme to help them settle into the new school environment. This may include campus tours, meeting teachers and classmates, and learning about the school's policies and expectations.

**Pricing**

Tuition fees for international schools in Thailand vary widely depending on the school's reputation, location, and facilities. Here's a general idea of what you can expect in terms of pricing:

**British curriculum schools:** Schools offering the British curriculum tend to have higher tuition fees. These fees typically range from mid to high levels, reflecting the comprehensive education and premium facilities provided. Parents can expect to pay a significant amount annually, depending on the grade level and the school's reputation. **THB 300,000 to THB 900,000**.

**American curriculum schools:** Tuition fees for American curriculum schools are generally comparable to British curriculum schools. These schools offer a variety of programmes, including Advanced Placement (AP) courses, which contribute to the overall cost. Annual fees can also vary widely based on the specific offerings and facilities of each school. **THB 300,000 to THB 850,000**.

**International baccalaureate schools:** IB schools in Thailand also have a range of tuition fees, often aligning with the rigorous academic standards and holistic development focus of the IB programme. The costs can be substantial, particularly for schools offering the full IB continuum from the Primary Years

Programme (PYP) to the Diploma Programme (DP). These fees will generally range from **THB 400,000 to THB 1,000,000**.

**Additional costs:** Besides tuition fees, there may be other costs to consider, such as enrolment fees, capital fees, and activity fees. Enrolment fees can range from **THB 50,000 to THB 300,000**, depending on the school. Capital fees, which contribute to the school's development and facilities, can also add to the overall cost. Activity fees for extra-curricular programmes, field trips, and sports can vary, so it's essential to factor these into your budget.

**Financial aid and scholarships:** Some international schools in Thailand offer financial aid and scholarships to support families with the cost of education. These may be based on academic merit, financial need, or specific talents in areas like sports or the arts. If you're concerned about the cost, it's worth exploring these options and discussing them with the school's admissions office.

**Conclusion**

Choosing the right international school in Thailand involves considering various factors, including curriculum options, enrolment processes, and pricing. By understanding the different curricula available, navigating the enrolment steps, and budgeting for tuition fees and additional costs, you can make an informed decision that best suits your child's educational needs. With top-notch international schools offering excellent academic programmes, state-of-the-art facilities, and a strong focus on holistic development, you can ensure your child receives a quality education while you enjoy the digital nomad lifestyle in the Land of Smiles.

## LOCAL SCHOOLS

Choosing the right educational environment for your children is crucial, especially when living as a digital nomad in Thailand.

While international schools offer a familiar curriculum and environment for expat families, local Thai schools can provide unique benefits and challenges. Enrolling your child in a local Thai school can offer a deep cultural immersion, language acquisition, and a different educational perspective. However, it may also come with language barriers and different teaching methodologies. Understanding the pros and cons of local Thai schools can help you make an informed decision that best suits your family's needs and aspirations.

## Local Thai Schools Analysis

### Pros of local Thai schools

**Cultural immersion:** One of the most significant advantages of enrolling your child in a local Thai school is the opportunity for cultural immersion. Your child will be exposed to Thai customs, traditions, and language daily, fostering a deep understanding and appreciation of the local culture. This immersion can lead to greater cultural sensitivity and adaptability, valuable skills in our increasingly globalised world.

**Language acquisition:** Attending a local Thai school can accelerate your child's proficiency in the Thai language. Being surrounded by native speakers in an academic setting provides a natural and effective way to learn the language. Bilingualism or even multilingualism can open doors to future opportunities and enhance cognitive abilities.

**Cost-effective:** Local Thai schools are generally more affordable than international schools. Tuition fees at local schools are usually a fraction of the cost of international schools, making them an attractive option for families on a budget. Additionally, other expenses, such as uniforms, books, and extra-curricular activities, tend to be more affordable.

**Community integration:** Enrolling your child in a local Thai school can help your family integrate into the local community.

Building relationships with Thai families, teachers, and students can create a support network and foster a sense of belonging. This integration can enrich your overall experience in Thailand and provide valuable insights into the local way of life.

**Diverse learning experiences:** Local Thai schools often focus on traditional subjects and methods, providing a different learning experience than international schools. This diversity can broaden your child's perspective and expose them to various teaching styles and approaches. Additionally, local schools may offer unique extra-curricular activities, such as traditional Thai dance, music, and martial arts.

**Cons of local Thai schools**

**Language barrier:** One of the most significant challenges of enrolling your child in a local Thai school is the language barrier. If your child is not already proficient in Thai, they may struggle to understand lessons and communicate effectively with teachers and classmates. This language barrier can impact their academic performance and social integration.

**Different teaching methods:** Local Thai schools often employ different teaching methods and curricula than international schools. The Thai education system tends to focus on rote learning and memorisation, which may not align with the teaching styles your child is accustomed to. This difference in approach can be challenging for some students and may require additional support and adaptation.

**Limited English support:** While some local Thai schools offer English language programmes, the level of support may not be sufficient for non-Thai speaking students. The quality and availability of English instruction can vary significantly between schools, making it essential to research and visit potential schools to assess their English support services.

**Academic standards:** The academic standards at local Thai schools can vary widely. While some schools maintain high standards and offer excellent education, others may have less rigorous curricula and resources. It's crucial to thoroughly research and visit schools to ensure they meet your expectations for academic quality and support.

**Social integration:** Integrating into a new social environment can be challenging, especially if your child is the only non-Thai student in their class. They may face difficulties making friends and fitting in, particularly if there are cultural differences and language barriers. Additional support and encouragement can help your child navigate these challenges and build meaningful relationships.

**Navigating the Thai education system**

Understanding the structure and expectations of the Thai education system can help you make an informed decision about enrolling your child in a local school. Here are some key aspects to consider:

**Curriculum:** The Thai education system follows a national curriculum set by the Ministry of Education. The curriculum includes core subjects such as Thai language, mathematics, science, social studies, and physical education. Some schools also offer additional subjects, such as English, art, and music. It's essential to review the curriculum and ensure it aligns with your child's educational needs and goals.

**School year and schedule:** The Thai school year typically runs from May to March, with two semesters. The first semester is from May to October, and the second semester is from November to March. Schools usually have a summer break in April and a mid-year break in October. The school day generally starts around 8:00am and ends around 3:00pm, with variations depending on the school and grade level.

**Class sizes:** Class sizes in local Thai schools can vary, but they tend to be larger than those in international schools. It's not uncommon for classes to have 30-40 students, impacting the level of individual attention and support your child receives. Visiting potential schools and observing classes can help you assess whether the class size is suitable for your child.

**Teacher qualifications:** Teacher qualifications and experience can vary significantly between local Thai schools. While some schools employ highly qualified and experienced teachers, others may have less stringent requirements. Researching the school's reputation and speaking with other parents can provide valuable insights into the quality of the teaching staff.

**Extra-curricular activities:** Local Thai schools often offer a range of extra-curricular activities, including sports, arts, and cultural programmes. Participating in these activities can help your child develop new skills, make friends, and build a sense of belonging. Reviewing the school's extra-curricular offerings can help determine whether they align with your child's interests and needs.

**Support services:** Assess the availability of support services, such as counselling, special education, and language assistance. These services can play a crucial role in your child's academic and social success, particularly if they face language barriers or other challenges. Visiting the school and speaking with staff can help you determine the level of support available.

**Making the decision**

Deciding whether to enrol your child in a local Thai school involves weighing the pros and cons and considering your family's unique circumstances. Here are some tips to help you make the best decision:

**Research and visit schools:** Thoroughly research potential schools and visit them in person to get a feel for the environment, facilities, and teaching staff. Speaking with other expat families and seeking their recommendations can also provide valuable insights.

**Consider your child's needs:** Assess your child's academic, social, and emotional needs and determine whether a local Thai school can meet them. Consider factors such as language proficiency, learning style, and adaptability when deciding.

**Evaluate your budget:** Compare the costs of local Thai schools and international schools to determine what fits within your budget. While local schools are more affordable, additional expenses, such as language tutoring or extra-curricular activities, may be necessary.

**Prepare for the transition:** If you decide to enrol your child in a local Thai school, prepare them by providing language support and familiarising them with Thai culture. Encourage open communication and provide a supportive environment to help them navigate challenges.

## Seamless Schooling Transition

Moving to a new country and enrolling your child in a local Thai school can be an exciting yet challenging experience. Ensuring a smooth transition involves addressing cultural integration and language support. Understanding local culture and language is essential for your child's academic success and overall wellbeing. In this section, we'll explore strategies for cultural integration, language support options, creating a supportive home environment, and connecting with other expat families. These insights will help your child thrive in their new environment and maximise their educational journey in Thailand.

### Cultural integration

Cultural integration is crucial for helping your child feel comfortable and accepted in their new school. Here are some strategies to foster cultural integration:

**Learn about Thai culture:** Before moving to Thailand, take the time to learn about Thai culture, customs, and traditions. Understanding the local way of life can help you and your child adapt more quickly. Resources such as books, documentaries, and online articles can provide valuable insights into Thai culture.

**Participate in cultural activities:** Encourage your child to participate in cultural activities both inside and outside of school. Many local Thai schools offer extra-curricular activities that focus on traditional arts, music, and dance. Engaging in these activities can help your child develop a deeper appreciation for Thai culture and build connections with their peers.

**Celebrate Thai holidays:** Celebrating Thai holidays and festivals is a fun way to immerse your family in the local culture. Events like Songkran (Thai New Year), Loy Krathong, and the Thai Lantern Festival offer unique opportunities to experience Thai traditions and customs. Participating in these celebrations can help your child feel more connected to their new environment.

**Build relationships with local families:** Forming friendships with local families can provide invaluable support and help your child feel more at home. Attend school events, join parent teacher associations, and participate in community activities to meet other families. Building a network of local friends can ease the transition and provide a sense of belonging.

**Encourage open communication:** Foster open communication with your child about their experiences and feelings. Encourage them to share their thoughts and concerns about adapting to the new culture. Providing a supportive environment where they feel comfortable discussing their challenges can help them navigate the adjustment process more effectively.

**Language support**

Language support is essential for helping your child succeed academically and socially in a local Thai school. Here are some language support options to consider:

**Language immersion programmes:** Some local Thai schools offer language immersion programmes to help non-Thai-speaking students learn the language more quickly. These programmes often include intensive Thai language classes, language labs, and one-on-one tutoring. Enrolling your child in a language immersion programme can provide the support they need to develop their language skills and succeed academically.

**Private language tutors:** Hiring a private language tutor can provide personalised support tailored to your child's needs. A tutor can work with your child on their language skills, help them understand their schoolwork, and provide additional practice outside school hours. Look for tutors with experience teaching Thai as a second language and a track record of success with students.

**Language exchange programmes:** Participating in language exchange programmes can be a fun and effective way for your child to improve their Thai language skills. These programmes pair students with native Thai speakers who want to learn English or another language. Through regular conversations and activities, both participants can practise their language skills and learn from each other.

**Online language courses:** There are numerous online language courses and apps available that can help your child learn Thai at their own pace. Platforms like **Duolingo, Rosetta Stone**, **BBC**

**Languages, FluentU, HelloTalk, Memrise, Italki, Mondly, Pimsleur, Tandem**, and **Babbel** offer interactive lessons, quizzes, and exercises to reinforce language learning. Incorporating online courses into your child's routine can provide additional practice and support.

**School support services:** Many local Thai schools offer support services for non-Thai-speaking students, such as language classes, tutoring, and resources for parents. Speak with the school's administration to understand the available support services and how they can benefit your child. Building a strong relationship with the school staff can also help you stay informed about your child's progress and any additional support they may need.

**Creating a supportive home environment**

In addition to external language support, creating a supportive home environment is crucial for your child's language development and cultural integration. Here are some tips for fostering a supportive home environment:

**Practice Thai at home:** Encourage your child to practice speaking Thai at home. Incorporate Thai into your daily routines, such as greeting each other in Thai, using Thai phrases during meals, and labelling household items with their Thai names. Practising Thai at home can help reinforce their language skills and build their confidence.

**Read Thai books and watch Thai media:** Expose your child to Thai books, movies, and TV shows to help them develop their language skills and gain a better understanding of Thai culture. Choose age-appropriate books and media that align with their interests to keep them engaged and motivated.

**Create a language-rich environment:** Surround your child with language-learning resources, such as Thai-English dictionaries, flashcards, and language apps. Encourage them to use these resources regularly to reinforce their learning and build their vocabulary.

**Be patient and supportive:** Learning a new language and adapting to a new culture can be challenging. Be patient and supportive as your child navigates this transition. Celebrate their progress, provide encouragement, and offer assistance when needed. Creating a positive and nurturing environment can help your child feel more confident and motivated to succeed.

**Connecting with other expat families**

Connecting with other expat families who have gone through similar experiences can provide valuable support and guidance. Here are some ways to connect with other expat families:

**Join expat communities:** Many cities in Thailand have active expat communities that organise events, meetups, and support groups. Joining these communities can help you meet other families, share experiences, and exchange advice on cultural integration and language support. Platforms like **InterNations, Meetup.com**, and local **Facebook Groups** for expats are excellent resources for finding these communities. Facebook, in particular, offers groups for everything: hiking, golf, reading, sewing, volunteering, cooking, language exchange, board games, soccer, etc.

**Attend school events:** Participating in school events, such as parent-teacher meetings, sports days, and cultural festivals, can help you connect with other expat families. Building relationships with other parents can provide a support network and help you navigate the challenges of adapting to a new culture and language.

**Share resources and tips:** Exchanging resources and tips with other expat families can help you discover effective strategies for cultural integration and language support. Share recommendations for language tutors, online courses, and cultural activities that have worked well for your family.

**Create playdates and social groups:** Organising playdates and social groups for your children can help them make friends and feel more at ease in their new environment. This can also provide opportunities for parents to connect and share experiences.

Engage in community service: Participating in community service projects can be a great way for your family to integrate into the local community and meet other expat families. Volunteering together can foster a sense of purpose and belonging while giving back to the community.

**Conclusion**

Enrolling your child in a local Thai school offers unique benefits like cultural immersion, language acquisition, and community integration while presenting challenges such as language barriers and different teaching methods. By carefully weighing the pros and cons and evaluating your child's needs, you can make an informed decision that best supports their educational journey. Cultural integration and language support ensure a smooth transition and a successful experience. Actively participating in cultural activities, providing language support, and creating a nurturing home environment can help your child thrive academically and socially. Connecting with other expat families and building a strong support network can further enhance your family's experience in Thailand.

## HOMESCHOOLING

For many digital nomad families, homeschooling presents a flexible and personalised educational option that fits seamlessly with a life of travel. Homeschooling allows you to tailor your child's education to their unique needs and interests, providing a rich and varied learning experience. Whether you're new to homeschooling or looking to enhance your current approach, having access to the right resources is crucial. In this section, we'll explore various online platforms and face-to-face groups that can support your homeschooling journey, ensuring your child receives a comprehensive and engaging education while you explore the world.

## Homeschooling 101

### Support network

Creating a robust support network is essential for a successful homeschooling experience. Here, we'll look at different ways to connect with other homeschooling families and access valuable resources.

### International communities

Joining international homeschooling communities can provide a wealth of support, resources, and camaraderie. These communities often offer online forums, social media groups, and local meetups where you can share experiences, seek advice, and find encouragement from like-minded families.

### Using the provided resources and testing

Accredited organisations and online platforms often offer a wide range of educational materials and testing options. Utilise these resources to ensure your child's education meets international standards and provides a thorough learning experience.

### Registering with your home country

To maintain compliance with educational regulations, consider registering your homeschooling programme with your home country. This can provide additional support and validation for your homeschooling efforts, ensuring your child's education is recognised globally.

**Utilising educational consultants**

Educational consultants can offer personalised advice and support tailored to your child's specific needs. These professionals can help you design a customised curriculum, recommend resources, and provide guidance on assessment and evaluation. By leveraging their expertise, you can enhance your homeschooling approach and ensure your child receives a high-quality education.

**Expat homeschooling organisations**

Finding the right organisation for you won't be easy as many are centred around cities/states, regions, countries, religions, and languages. However, there are some great ones tailored to expats, offering a range of support services and resources designed to meet the unique needs of families living abroad. These organisations can help you connect with other expat homeschooling families, provide access to international curricula, and offer guidance on legal requirements and best practices for homeschooling while living overseas.

- **Families in Global Transition (FIGT):** FIGT is a welcoming forum for globally mobile individuals, families, and those working with them. It offers valuable resources, networking opportunities, and support for homeschooling families living abroad. FIGT provides conferences, webinars, and a wealth of online resources to help expat families navigate the challenges and opportunities of homeschooling globally.
- **Worldschoolers:** This is a large online community that connects homeschooling families who travel the world. This organisation offers forums, social media groups, and local meetups where families can share experiences, resources, and support. Worldschoolers focuses on integrating travel and education, providing a platform for parents to exchange tips and ideas on creating

enriching learning experiences while exploring new cultures and destinations.

- **Global Homeschool Exchange (GHEX):** GHE connects homeschooling families across the globe, providing a platform for sharing resources, curriculum ideas, and support. This organisation offers online forums, webinars, and local events to help families build a strong homeschooling community. The Global Homeschool Exchange also provides guidance on international homeschooling regulations and best practices, ensuring that expat families can provide a high-quality education for their children no matter where they live.

**Online platforms**

The digital age has brought an abundance of online resources that make homeschooling more accessible and effective. Here are some top online platforms that offer a wealth of educational content and support for homeschooling families:

**Coursera:** Coursera partners with top universities and institutions worldwide to offer online courses and specialisations. The platform provides various subjects, from computer science and business to humanities and social sciences. While many courses are free, Coursera also offers paid options that provide certificates upon completion. Coursera's high-quality content and flexible learning options make it an excellent resource for homeschooling families looking to expand their child's education.

**Duolingo:** For families looking to incorporate language learning into their homeschooling curriculum, Duolingo is an excellent resource. This free language-learning platform offers interactive and gamified lessons in over 30 languages. Duolingo's bite-sized lessons and engaging format make language learning fun and accessible for students of all ages.

The platform also provides progress tracking and personalised learning paths to help your child achieve their language learning goals.

**EdX:** Like Coursera, EdX offers online courses from leading universities and institutions. The platform covers a wide array of subjects and provides both free and paid options. EdX's interactive and engaging courses are designed to foster critical thinking and problem-solving skills, making it a valuable resource for homeschooling families. Additionally, the platform's self-paced learning options allow your child to learn at their own pace and schedule.

**Khan Academy:** Khan Academy is a renowned online platform that provides free, high-quality educational resources for students of all ages. The platform covers a wide range of subjects, including mathematics, science, history, and art. With interactive lessons, practice exercises, and instructional videos, Khan Academy offers a comprehensive and engaging learning experience. The platform also provides personalised learning dashboards that allow you to track your child's progress and tailor their education to their specific needs.

**Outschool:** Outschool is an innovative online platform that offers live, interactive classes for students aged 3-18. The platform provides many subjects, including core academic subjects, arts, coding, and life skills. Outschool's small class sizes and experienced teachers create a personalised and engaging learning environment. The platform also offers flexible scheduling options, allowing you to choose classes that fit your child's interests and your family's travel schedule.

**Time4Learning:** Time4Learning is a comprehensive online curriculum provider that offers resources for students from Pre-K to Grade 12. The platform covers core subjects such as mathematics, language arts, science, and social studies. Time4Learning's interactive lessons, quizzes, and assessments provide a structured and engaging learning experience. The platform also offers flexible scheduling options and progress tracking to help you monitor your child's academic growth.

**Face-to-face groups**

While online platforms provide valuable resources and support, face-to-face interactions are also essential for a well-rounded homeschooling experience. Here are some ways to connect with other homeschooling families and groups in Thailand:

**Homeschooling co-ops:** Homeschooling co-ops are groups of families who come together to share resources, plan activities, and support each other in their homeschooling journey. Co-ops often organise group classes, field trips, and social events, providing opportunities for your child to interact with peers and learn in a collaborative environment. Joining a homeschooling co-op can help you build a supportive community and enhance your child's educational experience.

**Local meetups:** Platforms like **Meetup.com** and **Facebook Groups** are excellent resources for finding local homeschooling meetups and events. These gatherings allow your child to socialise with other homeschooled students and participate in group activities. Local meetups can also offer valuable support and advice from other homeschooling parents, helping you navigate the challenges and joys of homeschooling.

**Expat communities:** Many cities in Thailand have active expat communities that include homeschooling families. Joining expat groups, attending events, and participating in online forums can help you connect with other homeschooling families and share resources and experiences. Platforms like **InterNations** and local **Facebook Groups** for expats are great places to start.

**Educational workshops and classes:** Look for local educational workshops and classes that align with your child's interests and needs. Many cities in Thailand offer various extra-curricular activities, such as art classes, music lessons, and sports programmes. These face-to-face classes allow your child to learn new skills, interact with peers, and engage in hands-on learning experiences.

**Cultural and community events:** Participating in local cultural and community events can provide valuable learning opportunities and help your child connect with the local culture. Events such as festivals, museum exhibitions, and community fairs offer unique experiences that can enrich your homeschooling curriculum. Attending these events can also help you build connections with other families and expand your support network.

## Balancing Travel and Education

As a digital nomad family, one of the most significant challenges you may face is balancing the excitement of travel with the need for a consistent and effective education for your child. Homeschooling offers the flexibility to adapt learning to your travel schedule, but it requires careful planning and organisation. In this section, we'll explore strategies for balancing travel and education, ensuring your child receives a well-rounded education while you explore new destinations.

### Create a flexible schedule

One of the key advantages of homeschooling is the ability to create a flexible schedule that accommodates your travel plans. Here are some tips for developing a schedule that balances education and travel:

**Plan ahead:** Before embarking on your travels, create a tentative schedule outlining your itinerary and educational goals. Consider the duration of your stay, travel days, and planned activities. Having a rough plan in place can help you allocate time for both learning and exploration.

**Set realistic goals:** Establish realistic educational goals that align with your travel schedule. Break down larger objectives into smaller, manageable tasks that can be completed during travel. Flexibility is essential, so be prepared to adjust your goals as needed based on your travel experiences.

**Incorporate travel into learning:** Use your travel experiences as an opportunity to enhance your child's education. Incorporate lessons related to the places you visit, such as history, geography, and culture. For example, visiting historical sites, museums, and cultural festivals can provide valuable hands-on learning experiences that complement your homeschooling curriculum.

**Establish a routine:** While flexibility is crucial, having a daily or weekly routine can help provide structure and consistency. Dedicate specific times for academic work, such as mornings or evenings, and set aside time for exploration and leisure activities. A routine can help your child stay focused and maintain a sense of normalcy while travelling.

**Utilise travel time effectively**

Travel time, such as flights, train journeys, or long drives, can be used productively for educational activities. Here are some ways to make the most of travel time:

**Reading and audiobooks:** Carry a selection of books or e-books related to your child's curriculum and interests. Audiobooks are also a great option for long journeys, allowing your child to enjoy stories and learn while on the move.

**Educational apps and games:** Download educational apps and games that align with your child's learning objectives. These interactive tools can make learning fun and engaging, and they are easily accessible during travel.

**Journaling and creative writing:** Encourage your child to keep a travel journal where they can document their experiences, thoughts, and reflections. Creative writing assignments related to your travels can also be a valuable way to integrate education with exploration.

**Podcasts and documentaries:** Download educational podcasts and documentaries that cover topics relevant to your child's curriculum. These resources can provide valuable insights and knowledge while keeping your child entertained during travel.

**Maximise learning opportunities at destinations**

Each destination you visit offers unique learning opportunities that can enrich your homeschooling curriculum. Here are some strategies for maximising these opportunities:

**Research and plan:** Before arriving at a new destination, research the educational attractions and activities available. Look for museums, historical sites, cultural centres, and natural landmarks that align with your child's curriculum. Planning ahead can help you maximise your time and ensure a rich learning experience.

**Engage with locals:** Encourage your child to interact with local people and learn about their customs, traditions, and way of life. Language exchange programmes, cultural workshops, and community events can provide valuable insights and foster cross-cultural understanding.

**Participate in local classes and workshops:**Many destinations offer classes and workshops that can complement your child's homeschooling curriculum. Look for opportunities to learn traditional crafts, cooking, dance, or music. These hands-on experiences can provide a deeper understanding of the local culture and enhance your child's education.

**Explore nature and the environment:** Use your travels as an opportunity to explore different natural environments and ecosystems. Nature walks, wildlife tours, and environmental activities can provide valuable lessons in science, geography, and conservation.

**Stay organised and adaptable**

Staying organised and adaptable is essential for balancing travel and education. Here are some tips to help you stay on track:

**Keep records:** Maintain detailed records of your child's educational progress, including completed assignments, test scores, and learning milestones. Keeping organised records can help you track your child's development and ensure that they meet educational standards.

**Use technology:** Leverage technology to do research, study and stay organised. Use digital tools like **Google Drive, Evernote**, **Anki**, **ChatGPT**, **Slack**, **MindMeister**, and **Trello** to manage lesson plans, assignments, and resources. Online learning platforms and virtual classrooms can also provide valuable support and flexibility.

**Be adaptable:** Travel can be unpredictable, and plans may change unexpectedly. Embrace flexibility and be prepared to adapt your homeschooling schedule as needed. If travel delays or unexpected events disrupt your plans, look for alternative learning opportunities and adjust accordingly.

**Prioritise self-care:** Balancing travel and education can be demanding, so it's essential to prioritise self-care for you and your child. Ensure you get enough rest, stay healthy, and take breaks when needed. A healthy and well-rested mind is more productive and better equipped to handle the demands of homeschooling and travel.

**Involve your child in planning**

Involving your child in the planning process can help them take ownership of their education and stay motivated. Here are some ways to involve your child in planning:

**Set goals together:** Work with your child to set educational goals and objectives. Encourage them to take an active role in deciding what subjects and topics they want to explore. Collaborative goal setting can help your child feel more invested in their learning.

**Plan activities:** Involve your child in planning travel activities and educational excursions. Encourage them to research destinations, attractions, and events that interest them. Allowing your child to have a say in the planning process can make learning more engaging and enjoyable.

**Create a learning calendar:** Collaborate with your child to create a learning calendar that outlines their study schedule, travel plans, and extracurricular activities. This visual representation can help your child understand how their time is allocated and encourage them to take responsibility for managing their time effectively.

**Incorporate their interests:** Consider your child's interests and hobbies when planning lessons and activities. Tailoring the curriculum to include their passions can make learning more enjoyable and meaningful, keeping them motivated and engaged.

## Conclusion

Homeschooling as a digital nomad family offers unparalleled flexibility and the chance to tailor your child's education to their unique needs and interests. By leveraging online platforms, face-to-face groups, and careful planning, you can provide a comprehensive and engaging learning experience that supports your child's academic and personal growth. Create a flexible schedule, utilise travel time effectively, and maximise learning opportunities at each destination. Stay organised and adaptable, involve your child in the planning process, and prioritise self-care to enhance your homeschooling journey. Embrace the unique opportunities that travel offers and create a rich and varied educational experience that allows your child to thrive while exploring the world.

## EXTRA-CURRICULAR ACTIVITIES

Extra-curricular activities play a vital role in a child's development, offering opportunities to explore interests, build skills, and make new friends. For digital nomad families living in Thailand, these activities can provide a sense of stability and routine, enriching your child's educational journey. From sports and arts to music and cultural activities, Thailand offers a diverse range of options to keep your child engaged and entertained. In this section, we'll explore various extra-curricular activities available for kids in Thailand, helping you create a well-rounded and fulfilling experience for your child while embracing the digital nomad lifestyle.

### Activities for Kids

#### Sports

Sports are a fantastic way for children to stay active, develop teamwork skills, and build confidence. Thailand offers a wide range of sports activities for kids, catering to different interests and skill levels.

**Basketball:** Basketball is widely played in Thailand, and many schools and community centres have basketball courts and programmes for kids. Joining a basketball team can help your child develop teamwork skills, improve their physical fitness, and build lasting friendships.

**Football:** Football is hugely popular in Thailand, and many schools and local clubs offer football programmes for kids. Joining a football team can help your child develop their skills, make new friends, and stay physically fit. Look for local clubs or school teams that offer training sessions and matches for different age groups.

**Muay Thai:** Known as the national sport of Thailand, Muay Thai is a traditional martial art that combines striking and clinching techniques. Many gyms across the country offer Muay Thai classes for kids, focusing on fitness, discipline, and self-defence. Enrolling your child in Muay Thai classes can provide a unique cultural experience while promoting physical fitness and mental resilience.

**Racquet sports:** Tennis, badminton, and padel are all very popular in Thailand, with many clubs and sports centres offering lessons and facilities for kids. All of them help improve hand-eye coordination, agility, and overall fitness. Look for local clubs or coaches that offer group or private lessons tailored to your child's skill level.

**Swimming:** With its warm climate and beautiful beaches, Thailand is an ideal place for children to learn and enjoy swimming. Many schools, hotels, and sports centres offer swimming lessons for kids of all ages and skill levels. Swimming is not only a fun and healthy activity but also an essential life skill.

**Water sports:** Of course, Thailand is an excellent destination for children to explore various aquatic sports. Older kids can try scuba diving and discover the wonders of the deep. Surfing is another exciting option, with several surf schools offering lessons for beginners. Wakeboarding provides an adrenaline rush as children glide over the water's surface, while kite surfing combines elements of surfing and paragliding for a thrilling experience.

**Arts**

Engaging in artistic activities allows children to express their creativity, develop fine motor skills, and build confidence.

Thailand offers a variety of art programmes and classes that cater to different interests and abilities.

**Crafts and DIY projects:** Craft workshops and DIY projects are popular activities for kids in Thailand. These workshops often focus on traditional Thai crafts, such as making paper lanterns, weaving, and creating floral arrangements. Participating in craft workshops can provide a fun and educational experience, helping children learn about Thai culture and traditions.

**LEGO® Technic™:** Technic Lego offers a unique opportunity for children to develop their engineering and problem-solving skills. These sets include complex building projects that involve gears, motors, and other mechanical components. Many educational centres and clubs in Thailand offer Technic Lego workshops and classes where children can work on collaborative projects, learn about basic engineering principles, and enhance their creativity.

**Painting and drawing:** Many art studios and community centres in Thailand offer painting and drawing classes for kids. These classes provide a supportive environment for children to explore their artistic talents and develop their skills. Look for local art studios that offer classes in different mediums, such as watercolours, acrylics, and pastels.

**Sculpture and pottery:** Sculpture and pottery classes allow children to work with their hands and create three-dimensional art pieces. Many art centres and studios offer pottery classes for kids, where they can learn techniques like hand-building and wheel-throwing. These classes can help develop fine motor skills and foster a sense of accomplishment.

**Performing arts**

Performing arts activities, such as music, dance, and theatre, offer children opportunities to develop their talents, build confidence, and express themselves creatively. Thailand has a vibrant performing arts scene with numerous programmes and classes for kids.

**Dance:** Dance classes are a popular extra-curricular activity for kids in Thailand. From traditional Thai dance to ballet, hip-hop, and contemporary dance, there are numerous options to suit different interests and skill levels. Dance classes help children develop physical fitness, rhythm, and body awareness while providing a fun and social environment.

**Film and media:** Engaging in film and media activities can offer children a modern platform to express their creativity and storytelling abilities. Many institutions in Thailand provide workshops and courses in filmmaking, video editing, and animation. Kids can learn about scriptwriting, camera techniques, and post-production processes. These activities help children develop technical skills and creativity while understanding the power of visual storytelling.

**Music:** Learning to play a musical instrument or joining a choir can be a rewarding experience for children. Many music schools and private tutors in Thailand offer lessons in various instruments, such as piano, guitar, violin, and drums. Singing lessons and choir groups are also available for kids who enjoy vocal performance. Engaging in music activities can help improve concentration, coordination, and emotional expression.

**Theatre:** Participating in theatre productions and drama classes can help children build confidence, improve communication skills, and explore creativity. Many schools and community theatres in Thailand offer drama programmes for kids, where they can learn acting techniques, stagecraft, and improvisation. Theatre activities provide a supportive space for children to express themselves and collaborate with others.

**Other**

In addition to sports, arts, and performing arts, there are many other extra-curricular activities available for kids in Thailand that cater to a wide range of interests.

**Community service:** Encouraging your child to participate in community service projects can help them develop empathy, social responsibility, and a sense of purpose. Many schools and organisations in Thailand offer opportunities for kids to get involved in community service, such as volunteering at animal shelters, participating in beach cleanups, and supporting local charities. Community service activities can provide valuable life lessons and help children make a positive impact on their community.

**Language classes:** Learning a new language can be a valuable and enriching experience for children. Many language schools and tutors in Thailand offer classes in various languages, such as Thai, Mandarin, French, and Spanish. Language classes can help improve cognitive skills, cultural understanding, and communication abilities.

**Outdoor activities:** Thailand's diverse landscapes and natural beauty provide ample opportunities for outdoor activities and adventures. Activities like hiking, nature walks, and camping can help children develop a love for nature, build physical fitness, and learn valuable outdoor skills. Look for local outdoor clubs and organisations that offer guided activities and programmes for kids.

**STEM programmes:** Science, technology, engineering, and mathematics (STEM) programmes are increasingly popular for kids in Thailand. Many schools, community centres, and private organisations offer STEM classes and workshops that focus on subjects like coding, robotics, and environmental science. These programmes provide hands-on learning experiences that foster critical thinking, problem-solving, and innovation.

## Keeping Children Engaged and Socialised

For digital nomad families, keeping children engaged and socialised is essential for their well-being and development. While the nomadic lifestyle offers exciting experiences and learning opportunities, it can also present challenges in maintaining consistent social interactions and keeping kids actively involved. In this section, we'll explore strategies and activities to help your child stay engaged and socialised while living in Thailand. From joining local clubs and participating in community events to leveraging online platforms for virtual interactions, these tips will ensure your child remains connected, active, and happy in their new environment.

### Join local clubs and groups

Participating in local clubs and groups is an excellent way for your child to make friends and stay engaged. Here are some options to consider:

**Sports clubs:** Enrolling your child in local sports clubs, such as football, swimming, tennis, or martial arts, provides regular opportunities for physical activity and social interaction. Sports clubs often have training sessions, matches, and events that foster teamwork and camaraderie.

**Arts and crafts groups:** Many community centres and art studios in Thailand offer arts and crafts classes for children. These groups allow kids to explore their creativity, develop new skills, and interact with peers who share similar interests.

**Language exchange programmes:** Joining language exchange programmes can help your child improve their language skills while making new friends. These programmes pair students with native speakers, providing opportunities for conversation and cultural exchange.

**Cultural and hobby clubs:** Look for clubs focusing on specific hobbies or cultural activities, such as dance, music, chess, hiking, pet care, or gardening. These clubs offer a platform for your child to pursue their interests and connect with other children who share their passions.

**Engage in community events**

Community events provide valuable opportunities for social interaction and engagement. Here are some ways to get involved:

**Festivals and celebrations:** Thailand is known for its vibrant festivals and celebrations, such as Songkran, Loy Krathong, and the Thai Lantern Festival. Participating in these events allows your child to experience local culture, interact with the community, and create lasting memories.

**Local markets and fairs:** Attending local markets and fairs can be a fun and educational experience for your child. These events often feature local artisans, food stalls, and entertainment. They provide a unique opportunity for your child to interact with vendors, learn about local products and crafts, and engage in community activities.

**Workshops and classes:** Many cities in Thailand offer workshops and classes on various topics, from cooking and traditional crafts to science and technology. Enrolling your child in these classes can provide hands-on learning experiences and opportunities to meet new friends.

**Volunteering:** Volunteering for community service projects can help your child develop empathy, social responsibility, and a sense of purpose. Look for opportunities to get involved in beach cleanups, animal shelter support, fundraising, teaching English and local charity events.

**Attend social gatherings**

Attending social gatherings organised by expat communities, schools, or local groups can help your child build a network of friends and stay engaged. Here are some ideas:

**Expat meetups:** Many cities in Thailand have active expat communities that organise regular meetups and events. Joining these gatherings can help your family connect with other expats, share experiences, and build friendships.

**School events:** If your child is enrolled in a local or international school, attending school events such as sports days, cultural festivals, and parent-teacher meetings can provide opportunities for social interaction and community building.

**Playdates and family outings:** Organising playdates and family outings with other families in your community. Visiting parks, museums, and recreational centres can provide fun and engaging experiences for your child while fostering social connections.

**Neighbourhood gatherings:** Participate in neighbourhood gatherings or community block parties. These events often include games, activities, and shared meals, offering a relaxed environment for your child to interact with local families and make new friends. Getting involved in your neighbourhood can create a sense of belonging and strengthen community ties.

**Encourage independent activities**

Encouraging your child to pursue independent activities that interest them can help keep them engaged and motivated. Here are some suggestions:

**Reading and storytelling:** Encourage your child to read books that interest them or participate in storytelling sessions. Reading not only enhances cognitive skills but also provides a quiet and enjoyable way to stay engaged.

**Creative projects:** Provide materials for creative projects such as drawing, painting, building models, or crafting. Independent creative activities allow your child to explore their imagination and develop their skills.

**Online clubs and virtual classes:** Enrol your child in online clubs and virtual classes that align with their interests. These platforms offer various activities, from coding and robotics to art and music, providing opportunities for engagement and social interaction.

**Gardening and nature exploration:** Encourage your child to use gardening or nature exploration as a hobby. Gardening can teach them about plant life, responsibility, and the environment, while exploring nature can foster curiosity and a love for the outdoors. These activities provide hands-on learning experiences and can be both relaxing and educational.

**Maintain a balanced routine**

A balanced routine with educational, recreational, and social activities can help keep your child engaged and well-rounded. Here are some tips for creating a balanced routine:

**Set a schedule:** Establish a daily or weekly schedule with dedicated time for academic work, physical activity, creative projects, and social interactions. A consistent routine can provide structure and stability for your child.

**Incorporate variety:** Introduce a variety of activities to keep your child engaged and prevent boredom. Rotate between activities, such as sports, arts, and academic subjects, to maintain a well-rounded routine.

**Encourage breaks and downtime:** Ensure your child has time for breaks and relaxation. Downtime is essential for mental and emotional well-being, allowing your child to recharge and stay motivated. It's imperative that your child schedules healthy meals and physical education breaks into their day.

**Plan family activities:** Include regular family activities in your routine, such as game nights, outdoor excursions, or movie nights. These activities strengthen family bonds and provide opportunities for fun and relaxation. Family time is crucial for emotional support and creating lasting memories.

## Conclusion

Extra-curricular activities play a crucial role in a child's development, offering opportunities to explore interests, build skills, and make new friends. For digital nomad families living in Thailand, keeping children engaged and socialised is especially important. By joining local clubs and groups, participating in community events, and encouraging independent activities, you can provide a well-rounded and fulfilling experience for your child. Maintaining a balanced routine, staying connected with family and friends back home, and fostering social skills development further enhance your child's engagement and socialisation. Embrace the vibrant extra-curricular landscape and diverse opportunities available in Thailand to create a rich and supportive environment that allows your child to thrive while enjoying the digital nomad lifestyle.

# Resources

Congratulations on reaching the end of our guidebook. As you prepare to embark on your new life as a digital nomad in Thailand, this section will provide you with essential resources to ease your transition.

Why is this important? These resources will serve as the foundation of your new adventure. From government offices to help with legal matters, to clubs and groups for connecting with like-minded individuals, everything you need is here. Each link is a crucial step in navigating the diverse landscape of your new life.

Explore this carefully curated collection of resources. Use it to overcome challenges, seize opportunities, and enhance your experience in Thailand. The end of this book is the beginning of your journey. Wishing you safe travels and enriching experiences.

## Chapter 1: Living & Housing

### Accommodation

- Agoda – agoda.com
- Airbnb – airbnb.com
- Booking.com – booking.com
- The Ascott Limited – discoverasr.com/en/ascott-the-residence
- Oakwood – discoverasr.com//en/oakwood
- Somerset Serviced Residence – discoverasr.com/en/somerset-serviced-residence

### Appliances and Tools

- BnB Home – bnbhome.com
- DoHome – dohome.co.th

- Global House – globalhouse.co.th
- HomePro – homepro.co.th
- PowerBuy – powerbuy.co.th
- Tai Watsadu – taiwatsadu.com

**Co-Living Spaces**

- Alt Co-living – altcoliving.com
- Amy Village – facebook.com/amyvillagegardenresort
- El Pillax Lanta Resort – th.elpillaxlantaresort.com
- HOMA – homa.co
- Hub53 – hub53.com
- KoHub – kohub.org
- LITA – litabangkok.com
- lyf Sukhumvit Soi 8 – discoverasr.com/en/lyf/thailand/lyf-sukhumvit-8-bangkok
- Mimi & Jim: Co-Living Space – facebook.com/mimiandjim
- One Day – onedaybkk.com
- Roost Glamping – roostglamping.com
- Secret Place Hotels & Coworking – secret-place.info
- Selina – Serenity Rawai – selinahotels.com/phuket
- Socialer – socialer.thaiwanconcepts.com
- TAO HUB – taohub.asia
- The Cove – phangancove.com
- The Social Club – thesocialclub.asia
- The Yard Hostel – theyardhostel.com
- Ulima Co-living Hotel – delightful-garden-villa.bangkokshotels.com
- UnionSPACE – unionspace.co.th

**Facebook Groups**

- Thailand Expats – facebook.com/groups/ThailandExpats
- Bangkok Housing – facebook.com/groups/BangkokHousing

- Rental and Property – facebook.com/groups/RentalandProperty

**Forums and Info Sites**

- ASEAN Now – aseannow.com
- Couchsurfing – couchsurfing.com/places/asia/thailand
- Expat.com - Thailand Forum – expat.com/forum/viewforum.php?id=279
- Pattaya Addicts – pattaya-addicts.com
- Reddit - Thailand Travel – reddit.com/r/thailand
- Sawadee – sawadee.com
- TeakDoor – teakdoor.com
- Thaiger Talk – talk.thethaiger.com
- Thaifriendly – thaifriendly.com
- Thailand-247 – thailand-247.com
- ThailandQA – thailandqa.com

**Furniture Retailers**

- BnB Home – bnbhome.com
- Chanintr – chanintr.com
- Chic Republic – chicrepublicthai.com
- Euro Creations – eurocreations.co.th
- Facebook Marketplace – facebook.com/marketplace
- HomePro – homepro.co.th
- IKEA – ikea.com/th
- Index Living Mall – indexlivingmall.com
- NocNoc – nocnoc.com
- SB Design Square – sbdesignsquare.com

**Grocery and Supermarket Delivery Services**

- BigC – bigc.co.th
- HappyFresh – happyfresh.com
- Lotus's – lotuss.com
- Makro – makro.co.th

- Tops – tops.co.th
- Villa – shop.villamarket.com

**Meal Delivery Services**

- Aroi Dee – aroi-dee.com
- Eatigo – eatigo.com
- Foodpanda – foodpanda.co.th
- Grab – grab.com/th
- Robinhood – robinhood.co.th
- Wongnai – wongnai.com

**Meal Preparation Services**

- Absolute FitFood – absolutefitfood.com
- Getfresh – getfresh.co.th
- Healthy Meals – healthymealsthailand.com
- Nutrichef – nutrichef.net
- Paleo Robbie – paleorobbie.com
- Pure Prep – pureprep.co
- Vegan Crush – maricelsvegancrush.com

**Real Estate Sites**

- Conrad Properties – conradproperties.asia
- DDproperty – ddproperty.com
- Dot Property – dotproperty.co.th
- FazWaz – fazwaz.com
- Hipflat – hipflat.co.th
- Livinginsider – livinginsider.com
- PropertyScout – propertyscout.co.th
- RentHub – renthub.in.th
- Siam Real Estate – siamrealestate.com
- Thai Property – thaiproperty.com
- Thailand Property – thailand-property.com

# Chapter 2: Work & Productivity

## Chambers of Commerce

- American Chamber of Commerce (AmCham) – amchamthailand.com
- Australian Chamber of Commerce (AustCham) – austchamthailand.com
- British Chamber of Commerce (BCCT) – bccthai.com
- German-Thai Chamber of Commerce (GTCC) – thailand.ahk.de/en
- Thai-French Chamber of Commerce (FTCC) – francothaicc.com

## Co-Working Spaces

- Beachub – beachub.org
- CocoVille – cocovillephuket.com
- Garage Society – thegaragesociety.com
- Glowfish – glowfishworkplace.com
- Greenhouse – greenhousecowork.com
- Grind Time Coworking Space – grindtime.space
- HIVE Coworking Space – hiveworkspaces.com
- Hub53 – hub53.com
- Hubba-to – facebook.com/hubbato
- Kohub – kohub.org
- Koh Mak Campus – kohmakcampus.com
- Let's Work – letswork.co.th
- Oasis – oasis-kochang.com
- Punspace – punspace.com
- rPod Coworking Space – facebook.com/RPODyssey
- Seedspace – seedspace.co
- The Hive – thehive.co.th
- The Work Loft – theworkloft.com
- UnionSPACE – unionspace.co.th

- Yellow – yellowincubator.com

**Internet and Mobile Providers**

- 3BB – 3bb.co.th
- AIS – ais.th
- DTAC – dtac.co.th
- TOT – tot.co.th
- TrueMove – true.th

**Networking Apps**

- Bumble Bizz – bumble.com/bizz
- InterNations – internations.org
- Lunchclub – lunchclub.com
- Shapr – shapr.co
- Toastmasters Club – toastmasters.org

**Networking Events and Meetups**

- Eventbrite – .eventbrite.com
- Meetup – meetup.com
- Nomads.com – nomads.com

**Online Communities and Forums**

- Bangkok Entrepreneurs – facebook.com/groups/BangkokEntrepreneurs
- ExpatDen – facebook.com/groups/expatdenthailandcommunity
- LinkedIn – linkedin.com
- Lonely Planet – facebook.com/groups/lonelyplanettravellers
- Nomads.com– nomads.com
- Reddit – reddit.com
- Thailand Professionals – facebook.com/groups/thailandprofessionals

**Productivity Tools**

- Asana – asana.com
- Clockify – clockify.me
- Dropbox – dropbox.com
- Eisenhower Matrix – eisenhower.me/eisenhower-matrix
- Evernote – evernote.com
- Expensify – expensify.com
- Fiverr – fiverr.com
- Freelancer – freelancer.com
- Freedom – freedom.to
- Google Calendar – calendar.google.com
- Google Drive – google.com/drive
- Microsoft Outlook – outlook.live.com
- Notion – notion.so
- Pomodoro Technique – pomodorotechnique.com
- StayFocusd – stayfocusd.com
- Todoist – todoist.com
- Toggl – toggl.com
- Trello – trello.com
- Upwork – upwork.com
- World Time Buddy – worldtimebuddy.com

**Speed and Connectivity Tools**

- ADSL Thailand Speedtest – speedtest.adslthailand.com
- Speedtest by Ookla – speedtest.net
- WiFi Map – wifimap.io

**VPN Services**

- CyberGhost – cyberghostvpn.com
- ExpressVPN – expressvpn.com
- NordVPN – nordvpn.com

# Chapter 3: Finance & Budgeting

## Bangkok Shopping

- CentralWorld –centralworld.co.th
- Chatuchak Weekend Market – chatuchakmarket.org
- EmQuartier – emquartier.co.th
- EmSphere – emsphere.co.th
- Emporium – emporium.co.th
- Fortune Town – fortunetown.co.th
- Gateway Ekamai – facebook.com/gatewayekamai
- Gaysorn Centre – gaysornvillage.com
- ICONSIAM – iconsiam.com
- MBK Center – wmbk-center.co.th
- Siam Center – siamcenter.co.th
- Siam Paragon – siamparagon.co.th
- Siam Square – siam-square.com
- The Platinum Shopping Mall – platinumfashionmall.com

## Banks

- Bangkok Bank (BBL) – bangkokbank.com
- Kasikorn Bank (KBank) – kasikornbank.com
- Krungthai Bank (KTB) – krunghtai.com
- Siam Commercial Bank (SCB) – scb.co.th

## Cinemas

- Bangkok Screening Room – bkksr.com
- Cinema Oasis – cinemaoasis.com
- House Samyan – housesamyan.com
- Lido Connect – lidoconnect.com
- Major Cineplex – majorcineplex.com
- SF Cinema – sfcinema.com
- The Friese-Greene Club – fgc.in.th

**Driving**

- Avis – avis.com
- Budget – budget.com
- Department of Land Transport (DLT) – dlt.go.th
- Eazy Car – easycar.co
- Hertz – hertz.com
- International Driving Permit (IDP) – internationaldrivingpermit.org
- Sixt – sixt.com

**Educational Resources**

- Coursera – coursera.org
- edX – edx.org
- Udemy – udemy.com

**Financial Tools**

- H&R Block – hrblock.com
- TaxAct – taxact.com
- TurboTax – turbotax.intuit.com

**Nationwide Shopping Mall Chains**

- BigC – bigc.co.th
- Central – central.co.th
- Lotus's (formerly Tesco Lotus) – lotuss.com
- Robinson – central.co.th/en/shop/robinson
- Terminal 21 – terminal21.co.th
- The Mall – themall.co.th

**Nightlife**

- Bangkok Muay Thai at Rajadamnern – rajadamnern.com
- Pattaya Alcazar Cabaret Show – alcazarshow.pattaya-tickets.com
- Phuket FantaSea Show – phuket-fantasea.com

**Online Resources**

- Thai Revenue Department – rd.go.th
- TripAdvisor – tripadvisor.com
- Google Reviews (for specific reviews, use Google Maps search) – maps.google.com
- Reddit Tax Forums – reddit.com/r/tax
- Expat Forum Thailand – expatforum.com

**Payment Apps**

- PayPal – paypal.com
- Revolut – revolut.com
- Revolut Premium – revolut.com/revolut-premium
- Skrill – skrill.com
- Wise – wise.com

**Remittance Services**

- MoneyGram – moneygram.com
- Western Union – westernunion.com
- WorldRemit – worldremit.com
- xe.com – xe.com

**Ridesharing**

- Bolt – bolt.eu/th
- Grab – grab.com/th
- inDrive – indrive.com/en-th

**Spas**

- Asia Herb Massage & Spa – asiaherb.asia
- Divana Spa – divana-dvn.com
- Health Land Massage – healthlandspa.com
- Let's Relax Spa – letsrelaxspa.com
- Oasis Spa – oasisspa.net

### Tax

- Thai Revenue Department – rd.go.th

### Tax Advisors

- Bright!Tax – brighttax.com
- Double M – double-m.co
- Expat Tax Online – expattaxonline.com
- Expat Tax Thailand – expattaxthailand.com
- Forvis Mazars Thailand – forvismazars.com/th
- Greenback Expat Tax Services – greenbacktaxservices.com
- Network Advisory Team (NAT) – thaitaxaccounting.com
- Siam Legal – siam-legal.com
- Tax Samaritan – taxsamaritan.com

### Transport

- BRT Rapid Transit Bus – transitbangkok.com
- BTS Skytrain – bts.co.th
- MRT Subway – bangkokmetro.co.th

## Chapter 4: Visa & Legal Requirements

### Government Forms

- 90-Day Reporting (TM.47 Form) – immigration.go.th
- Medical Health Certificate (work permit) – doe.go.th
- Notification of Residence (TM.30) – immigration.go.th
- Thai Visa Extension Forms (TM.7) – immigration.go.th

### Government Sites

- Ministry of Education – moe.go.th
- Ministry of Labour – mol.go.th
- Royal Thai Embassy – thaiembassy.org
- Smart Visa – smart-visa.boi.go.th

- Thailand Elite Visa – thailandprivilege.co.th
- Thai Immigration Bureau – immigration.go.th

## Chapter 5: Health & Wellness

### Fitness Centres and Gyms

- Absolute You – absoluteboutiquefitness.co
- Anytime Fitness – anytimefitness.co.th
- BASE – basebangkok.com
- Carve House – carvehouse.co
- Fitness First – fitnessfirst.com/th
- Fitwhey Gym – fitwhey.com/gym
- Jetts 24 Hour Fitness – jetts.co.th
- The Edge Boutique Gym – theedgeboutiquegym.com
- True Fitness – truefitness.com
- Virgin Active – virginactive.co.th
- WeFitness Society – wefitnesssociety.com

### Healthcare and Insurance

- Aetna International – aetnainternational.com
- Allianz Ayudhya – azay.co.th
- Allianz Worldwide Care – allianz.co.th
- AXA Thailand – axa.co.th
- Bupa Global Health Insurance – bupaglobal.com
- Cigna Global Health Insurance – cignaglobal.com
- GeoBlue – geobluetravelinsurance.com
- Luma Health Insurance – lumahealth.com
- Pacific Cross Health Insurance – pacificcrosshealth.com
- Thai Health Insurance – thaihealth.co.th

### Meditation Retreats

- Dipabhāvan Meditation Center – dipabhavan.org
- Suan Sati – suansati.com
- Wat Pa Tam Wua Forest Monastery – wattamwua.com

- Wat Suan Mokkh – suanmokkh-idh.org

**Specialised Fitness Centres**

- Alpha Health Club – alphahealthclub.com
- Bangtao MuayThai and MMA – bangtaomuaythai.com
- CrossFit Chiang Mai (CFCNX) – cfcnxfitness.com
- Delta 55 Fitness – delta55fit.com
- Elite Gym and Fitness – elitegymsamui.com
- F45 and FS8 – f45training.com
- GoGym – facebook.com/gogymchiangmai.com
- Gym Bangarang – gymbangarang.com
- Krabi Fitness – facebook.com/thegymkrabi
- Krabi Lion Muay Thai – facebook.com/p/Krabi-Lion-Muay-Thai-61552646480668
- Lanta Gym Muay Thai – lantagym.com
- Lanta Sport Resort – facebook.com/Lantasport
- Padel Phuket – padel-phuket.com
- Siam Sports Pro – siamsports.pro
- Sly Black Tennis Academy – slyblacktennis.com
- Superpro Samui – superprosamui.com
- Thanyapura – thanyapura.com
- The LAB – thelabbangkok.com
- True Arena Hua Hin – truearenahuahin.com

**Vaccination and Health Information**

- Center for Disease Control and Prevention (CDC) Travel – cdc.gov/travel
- Thai Ministry of Public Health – moph.go.th
- World Health Organization (WHO) Travel Advice – who.int/travel-advice

**Wellness Centres**

- Absolute Sanctuary – absolutesanctuary.com
- Chiva-Som – chivasom.com

- Kamalaya – kamalaya.com
- Phuket Cleanse – phuketcleanse.com
- The Sanctuary – thesanctuarythailand.com

### Yoga Studios

- Lanta Yoga – lanta-yoga.com
- The Yoga Tree – facebook.com/yogatreechiangmai
- Yoga Balance – yogabalancethailand.com
- Yoga Elements Studio – yogaelements.com
- Yoga Republic – yogarepublic.co

## Chapter 6: Cultural Insights & Learning Thai

### Event Info Sites

- ASEAN Now – aseannow.com
- Bangkok Post – bangkokpost.com
- BK Magazine – bk.asia-city.com
- Eventbrite – eventbrite.com
- Lifestyle+Travel – lifestyleandtravel.com
- Local Facebook Groups – facebook.com/groups
- Phuket.Net – phuket.net
- Thailand Convention & Exhibition Bureau (TCEB) – businesseventsthailand.com
- Thailand Event Guide – thailandeventguide.com
- Thailand Magazine – thailandmagazine.com
- TAT Newsroom – tatnews.org
- Time Out Magazine – timeout.com/bangkok
- Tourism Authority of Thailand (TAT) – tourismthailand.org

### Language Schools

- ALA Language Center – alalanguage.com
- Duke Language School – dukelanguage.com
- Thai Language Hut – thailanguagehut.com

### Learning Thai

- ChatGPT – openai.com/chatgpt
- Coursera – coursera.org
- Duolingo – duolingo.com
- Google Translate – translate.google.com
- Italki – italki.com
- Learn Thai with Mod – learnthaiwithmod.com
- Mondly – mondly.com
- Pimsleur – pimsleur.com
- Preply – preply.com
- ThaiPod101 – thaipod101.com
- Udemy – udemy.com

## Chapter 7: Travel & Exploration

### ATV Adventures

- 8 Adventures – 8adventures.com
- ATV Pattaya Jungle Adventures – atvpattaya.com
- Bon Voyage Thailand – bonvoyagethailand.com
- CRD Adventure – crdadventure.com
- Pattaya Off-Road Safari – pattayaoffroadsafari.com
- Phuket ATV Tour – phuketatvtour.com
- Phuket Paradise Trip ATV Adventure – phuketatvtour.net
- River Kwai Village – riverkwaivillagehotel.com
- SSJ ATV PAI – facebook.com/ATVSSJ
- X-Centre Chiang Mai – chiangmaixcentre.com

### Budget Airlines

- AirAsia – airasia.com
- Nok Air – nokair.com
- Thai Lion Air – lionairthai.com

### Cycling Tours

- Bangkok Bike Tours – gobangkokbiketours.com

- Chiang Mai Bicycle – chiangmaibicycle.com
- Grasshopper Adventures – grasshopperadventures.com
- SpiceRoads Cycling – spiceroads.com

**Ferries and Speedboats**

- Andaman Wave Master – andamanwavemaster.com
- Klook – klook.com
- Lomprayah – lomprayah.com
- Phuket Ferry – phuketferry.com
- Seatran – seatrandiscovery.com
- Tigerline Ferry – tigerlineferry.com

**Forums and Info Sites**

- Culture Trip – theculturetrip.com
- Fodor's Travel – fodors.com/world/asia/thailand
- Lonely Planet – facebook.com/groups/lonelyplanettravellers
- Nomads.com – nomads.com
- Rough Guides – roughguides.com
- Thailand: Off the Beaten Path – facebook.com/groups/ThailandOffTheBeatenPath
- Thailand Tourism Directory – tourismthailand.org
- Travelfish – travelfish.org
- Tripadvisor Thailand Forum – tripadvisor.com/ShowForum-g293915-i3686-Thailand.html
- Wikitravel – wikitravel.org

**Hostels**

- Booking.com – booking.com
- Hostelworld – hostelworld.com
- Tripadvisor – tripadvisor.com

## House and Pet Sitting

- HouseCarers – housecarers.com
- Official Thailand House and Pet Sitting Facebook Group – facebook.com
- TrustedHousesitters – trustedhousesitters.com

## Kiteboarding, Wakeboarding, and Cable Skiing

- Club Loong Chat – cubloongchat.com
- ION Club – ion-club.net
- KBA Pattaya – kbapattaya.com
- KBA Phuket – kiteboardingasia.com
- Kite School Pranburi – gokite.asia
- Kite Thailand – kitethailand.com
- Kiteboarding Asia – kiteboardingasia.com
- Kite Zone Thailand – kitesurfthailand.com
- Phuket Wake Park – phuketwakepark.com
- Siam Pro Kite – siamprokite.com
- Surfspot – surfspothailand.com
- Thai Wake Park – thaiwakepark.com

## Local Transport

- Airport Rail Link Bangkok – srtet.co.th
- Bangkok Mass Transit Authority (BMTA) – bmta.co.th
- Bangkok MRT – bemplc.co.th
- BTS Skytrain – bts.co.th
- State Railway of Thailand (SRT) – railway.co.th

## Minivans and Buses

- 12Go Asia – 12go.asia
- Bell Travel Service – belltravelservice.com
- Bus Online Ticket – busonlineticket.co.th
- Green Bus – greenbusthailand.com
- Nakhonchai Air – nakhonchaiair.com

- Thai to Siam – thai2siam.com

**Rafting**

- Eco Tours Chiang Mai – ecotourschiangmai.com
- Khao Sok Lake – khaosoklake.com
- Khao Sok River Lodge – khaosokriverlodge.com
- Pai Tubing – paitipsytubing.com
- Siam River Adventures – siamrivers.com
- Thai Adventure Rafting – thairafting.com

**Ridesharing Apps**

- Bolt – bolt.eu
- Grab – grab.com
- inDrive – indrive.com

**Rock Climbing**

- 57 Hours – 57hours.com/adventure/lopburi-rock-climbing
- Basecamp Tonsai – basecamptonsai.com
- Chiang Mai Rock Climbing Adventures (CMRCA) – thailandclimbing.com
- King Climbers – king-climbers.com
- Koh Yao Travel – kohyaotravel.com
- Krabi Rock Climbing – krabirockclimbing.com
- Real Rocks Climbing School – realrocksclimbing.com
- Rock Shop – justclimbhailand.com
- Spider Monkey Climbing – spidermonkeyphiphi.com
- Thai Climbing Development Association (TCDA) – thaiclimbassociaiton.org

**Scuba Diving**

- Big Blue Diving – bigbluediving.com
- Blue Planet Divers – blueplanetdivers.net
- Crystal Dive – crystaldive.com

- Scuba Shack – scubashcakkohtao.com
- Sea Bees Diving – sea-bees.com
- Sea Dragon Dive Center – seadragondivecenter.com
- Similan Diving Safaris – similan-diving-safaris.com
- The Junk Liveaboards – thejunk.com

**Sightseeing Spots**

- Ban Chiang National Museum – thailandtourismdirectory.go.th
- Chi Lin Nunnery – nunnerychi.org
- Gardens by the Bay – gardensbythebay.com.sg
- Hellfire Pass – hellfire-pass.com
- Hong Kong Museum of History – hk.history.museum
- Sanctuary of Truth – sanctuaryoftruth.com
- S.E.A. Aquarium – rwsentosa.com/en/attractions/sea-aquarium
- Sentosa Island – sentosa.com.sg
- Tuol Sleng Genocide Museum – tuolsleng.gov.kh
- Universal Studios Singapore – rwsentosa.com/en/attractions/universal-studios-singapore
- War Remnants Museum – baotangchungtichchientranh.vn

**Trekking Tours**

- Chiang Mai Trekking – chiangmai-trekking.com
- G Adventures – gadventures.com
- Intrepid Travel – intrepidtravel.com
- Thailand Hilltribe Holidays – thailandhilltribeholidays.com
- Trek Thailand – trekthailand.net

**Ziplining**

- Flight of the Gibbon – flightofthegibbon.com

- Koh Phangan Jungle Flight – facebook.com/p/Jungle-Flight-Adventure-Park-Koh-Phangan-61556008730921/?_rdr
- Tarzan Adventure Pattaya – facebook.com/TarzanZipline
- Tarzan Adventure Phuket – facebook.com/tarzanphuket
- Tree Top Adventure Park – treetopadventurepark.com

## Chapter 8: Food & Dining

### Bib Gourmand Recommendations

- Go Benz Rice Porridge – facebook.com/GoBenzPhuket
- Huen Phen – michelin.com/th/restaurants/huen-phen
- Jay Oh – facebook.com/Jaekoy.JayOh
- Khao Soi Khun Yai – michelin.com/th/restaurants/khao-soi-khun-yai
- Kiti Panit – kitipanit.com
- Krua Tungtong – kruatungtong.com
- Likhit Kai Yang – michelin.com/th/restaurants/likhit-kai-yang
- Lock Tien – facebook.com/pages/ร้านอาหารลกเที้ยน-Lock-Tien-Phuket-Phuket
- Mor Mu Dong – michelin.com/th/restaurants/mor-mu-dong
- Prachak Roasted Duck – facebook.com/PrachakRoastedDuck
- Raya – rayarestaurantphuket.com
- Rung Rueang Pork Noodle – michelin.com/th/restaurants/rung-rueang-pork-noodle
- Soei – @soei_restaurnat (Instagram)
- Suan Pak – facebook.com/Kuasaunpak
- Tong Tem Toh – facebook.com/TonTemToh

**Cooking Schools**

- Baipai Thai Cooking School – baipai.com
- Blue Elephant Cooking School – blueelephant.com
- Chiang Mai Thai Cookery School – facebook.comCMTCS1993
- Phuket Thai Cooking Academy – phuketthaicookingacademy.com
- Smart Cook Thai Cookery School – facebook.com/SmartCookKrabi
- Sompong Thai Cooking School – facebook.com/sompongthaicookingschool
- Thai Farm Cooking School – thaifarmcooking.net
- Ya's Cookery School – yascookeryschool.com

**Michelin-Starred Restaurants**

- 80/20 – 8020bkk.com
- Aksorn – aksornbkk.com
- Baan Tepa – baantepabkk.com
- Blue by Alain Ducasse – blue-alainducasse.com
- Cadence by Dan Bark – canence-danbark.com
- Canvas – canvasbangkok.com
- Chef's Table – lebua.com/restaurants/chefs-table
- Chim by Siam Wisdom – chimbysiamwisdombkk.com
- Côte by Mauro Colagreco – cotebkk.com
- Elements, Inspired by Ciel Bleu – okurabangkok.com
- Gaa – gaabkk.com
- Haoma – haoma.dk
- IGNIV – ignivbangkok.com
- Inddee – inddeebkk.com
- Jay Fai – @jayfaibangkok (Instagram)
- Khao (Vadhana) – khaorestaurants.com
- Le Du – ledubkk.com

- Le Normandie – mandarinoriental.com/en/bangkok/chao-phraya-river/dine/le-normandie
- Maison Dunand – maisondunand.com
- Mezzaluna – lebua.com/restaurants/mezzaluna
- Mia – miarestaurantbkk.com
- Michelin Guide – guide.michelin.com/th/en
- Nahm – comohotels.com/thailand/como-metropolitan-bangkok/dining/nahm
- Nawa – @nawa.thaicuisine (Instagram)
- Potong – restaurantpotong.com
- PRU – prurestaurant.com
- R-Haan – r-haan.com
- Resonance – resonancebkk.com
- Samrub Samrub Thai – samrubsamrubthai.com
- Saneh Jaan – sanehjaan.com
- Sorn – sornfinesouthern.com
- Signature – signaturebangkok.com
- Suan Thip – @suanthip (Instagram)
- Sühring – restaurantsuhring.com
- Sushi Masato – sushimasato.com
- Yu Ting Yuan – fourseasons.com/bangkok/dining/restaurants/yu-ting-yuan

**Online Grocery Shopping**

- 7 Delivery – 7eleven.co.th
- Bangkok Grocery – bangkokgrocery.com
- Big C – bigc.co.th
- CP Freshmart – cpfreshmartshop.com
- Foodpanda – foodpanda.co.th
- Grab – grab.com/th
- Gourmet Market – gourmetmarketthailand.com
- HappyFresh – happyfresh.com

- Lazada – lazada.co.th
- Lotus's (formerly Tesco Lotus) – lotuss.com
- Makro – siammakro.co.th
- Shopee – shopee.co.th
- Tops Market – tops.co.th
- We Fresh – wefreshthailand.com

**Resources and Apps**

- AllergyEats – allergyeats.com
- Food Allergy Translation App – askallergies.com
- Google Translate – translate.google.com
- HappyCow – happycow.net
- Spokin – spokin.com
- Vegan-Friendly Destinations in Thailand – thailandholidaygroup.com
- Vegan Passport – vegansociety.com
- Vegetarian and Vegan Guide to Thailand – gotothailand.com

**Restaurants**

- Baan Khanitha – baan-khanitha.com
- Best Beef – facebook.com/BestBeefBarBQ
- Blue Elephant – blueelephant.com
- Khao Soi Khun Yai – @khaosoikhunying (Instagram)
- Laab Ubon – facebook.com/Laabubon
- Mae Varee – @maevareeoffical (Instagram)
- Nai Mong Hoi Tod – facebook.com/hoitod539
- Pe Aor – No link available
- Sala Rattanakosin Eatery and Bar – salahospitality.com
- Somtum Der – somtumder.com
- Supanniga Eating Room – supannigaeatingroom.com
- Thip Samai – thipsamai.com

### Socially Conscious Cooking Schools

- Community-Based Tourism Cooking Classes (Phuket) – thaicookingclassphuket.com
- Cooking with Poo and Friends – cookingwithpoo.com
- Thai Akha Kitchen – thaiakhakitchen.com

### Specialty Stores

- BritiShop – britishop.com
- Gourmet Market – gourmetmarketthailand.com
- iHerb – th.iherb.com
- Ninelife – ninelife.com
- Panama Market – panama.taplink.ws
- Rimping Supermarket – rimping.com
- Sunshine Market – sunshinemarket.co.th
- Villa Market – villamarket.com

## Chapter 9: Community & Social Life

### Cafés and Coffee Shops

- Akha Ama Coffee – akhaama.com
- Bookhemian – bookhemianphuket.wixsite.com/bookhemian
- Casa Lapin – casalapin.com
- Coffee Lab – coffeelabasia.com
- Nine One Coffee – nineonecoffee.co
- Project Artisan – theprojectartisan.com
- Ristr8to – ristr8to-coffee-chiangmai.com
- Rocket Coffee Bar – rocketbkk.com
- Roots Coffee – rootsbkk.com
- The Barisotel by The Baristro – facebook.com/thebarisotelbythebaristro
- The COMMONS – thecommonsbkk.com
- Umai Teahouse – facebook.com/people/UMAITeahouse/100094247930683

**Co-Working Spaces**

- Garage Society – thegaragesociety.com
- Greenhouse – greenhouse.co
- HOMA – homa.co
- Let's Work – letswork.co.th
- Punspace – punspace.com
- Regus Phuket – regus.com
- Starwork – starworkchiangmai.com
- The Hive – thehive.co.th
- The Work Loft – theworkloft.com
- WOLF Co-working Space – thewolfofficial.com
- UnionSPACE – unionspace.co.th
- Yellow Co-working – yellowincubator.com

**Community-Based Tourism**

- Baan Talae Nok Village – andamandiscoveries.com/homestay-experience-ban-talae-nok
- Koh Yao Noi Community-Based Tourism Club – kohyao-cbt.net
- Mae Kampong Village – fanclubthailand.co.uk/enjoy-northern-thai-hospitality-in-mae-kampong

**Conferences and Seminars**

- D.NX Conference – dnxfestival.com
- Dragonfly – dragonfly360.co
- Nomad Summit – nomadsummit.com
- RISE Conference – riseconf.com
- Tech in Asia – techinasia.com
- Techsauce Global Summit – summit.techsauce.co
- Thailand Wellness Summit – facebook.com/thailandwellnessasiasummittwas
- Wild Digital – wild-digital.com

**Cultural and Arts Events**

- Bangkok Art Biennale – bkkartbiennale.com
- Bangkok International Festival of Dance & Music – bangkokfestivals.com
- Chiang Mai Flower Festival – facebook.com/charmingchiangmaiflowerfestival
- Pattaya International Music Festival – pattaya.go.th
- Wonderfruit – wonderfruit.co

**Event Info and Tickets**

- Eventbrite – eventbrite.com
- Eventpop – eventpop.me
- Facebook Events – facebook.com/events
- Meetup – meetup.com
- ThaiTicketMajor – thaiticketmajor.com
- Ticketmelon – ticketmelon.com

**Facebook Groups**

- Bangkok Digital Nomads – facebook.com/groups/bangkokdigitalnomads
- Chiang Mai Digital Nomads – facebook.com/groups/chiangmaidigitalnomads
- Digital Nomads in Thailand – facebook.com/groups/digitalnomadthailand
- Phuket Digital Nomads – facebook.com/groups/phuketdigitalnomads
- Thailand Expats – facebook.com/groups/ThailandExpats

**Fitness Events**

- Bangkok Marathon – bkkmarathon.org
- Ironman Thailand – ironman.com
- Laguna Phuket Triathlon – lagunaphukettri.com
- Spartan Race Thailand – spartanrace.co.th

- Thailand Yoga Festival – asiawellness.net

**Homestays**

- Cultural Homestay International (CHI) – chinet.org
- Global Homestay Network (GHN) – globalhomestaynetwork.com
- XploreAsia Thailand Homestay Program – xploreasia.org/amazing-thailand-homestay-program/

**Language Immersion**

- AAA Thai Language School – aaathai.school
- Alpha Language School – lang-school.com
- AUA Language Center – auathailand.org
- Chiang mai University Language Institute – li.cmu.ac.th
- Duke Language School – dukelanguage.com
- Learn Thai from a White Guy – learnthaifromawhiteguy.com
- Mind Your Language – mindyourlanguagethailand.com
- Oriental Bamboo Language School – orientalbamboo.ac.th
- Phuket Thai Language School – phuket-languageschool.com
- PickupThai Podcast – pickup-thai.com
- PLC Language School Pattaya – plcpattaya.com
- Pro Language School – prolanguage.co.th
- Samui Language School – samuilanguageschool.org
- Sirada – The Learning Center – learningcentre.co.th
- Speak Thai Easy – speakthaieasy.com
- Sumaa Language and Culture Institute – sumaa.net
- The Language Centre Hua Hin – thailanguagecentre.org
- Thai Language Hut – thailanguagehut.com
- Thaipod101 – thaipod101.com

**Local Activities & Clubs**

- Bangkok Community Theatre – bangkokcommunitytheatre.com
- Bangkok Hikers – bangkokhiking.com
- Bangkok Runners – meetup.com/bangkok-runners
- Bangkok Lions Rugby Club – bkklionsrugby.org
- Chiang Mai Art Conversation – cac-art.info
- Chiang Mai Cycling Club – 55cyclingclub.com
- Chiang Mai Hiking Club – chiangmaihiking.wordpress.com
- Hash House Harriers – hashhouseharriersthailand.com
- InterNations – internations.org
- Phuket Hiking Club – facebook.com/PhuketHikingClub
- Phuket Yacht Club – phuketyachtclub.com
- Royal Varuna Yacht Club – varuna.org
- Samui Tennis Club – facebook.com/Samuitennisaddiction
- Toastmasters International – toastmasters.org

**Meetups and Social Events**

- Bangkok Entrepreneurs – bangkok-entrepreneurs.com
- BNOW.org – bnow.org
- Chiang Mai Expats Club – chiangmaiexpatsclub.com
- Garage Society – thegaragesociety.com
- Let's Work – letswork.co.th
- Nomad Coffee Club – facebook.com/groups/nomadcoffee
- Phuket Startup Businesses – phuketstartupbusinesses.com
- Phuket Expats Club – phuketexpatsclub.wordpress.com
- The Hive – thehive.co.th
- WeWork – wework.com

**Platforms**

- Discord – discord.com
- Expat.com – expat.com
- Facebook Events – facebook.com/events
- Meetup.com – meetup.com
- Nomads.com – nomads.com
- Nomad City – nomadcity.org
- Nomad Soulmates – nomad-soulmates.com
- Nomad Summit – nomadsummit.com
- Reddit – reddit.com
- Slack – slack.com
- WebWork Travel – webworktravel.com
- WiFI Trible – wifitribe.com

**Student Exchanges**

- AFS Intercultural Programs – afs.org
- Rotary Youth Exchange – rotary.org/en/our-programs/youth-exchanges
- Youth For Understanding (YFU) – yfu.org

**Technology and Innovation Events**

- ASEAN Angel Alliance Summit (AAAS) – facebook.com/ASEANAngels
- Echelon Thailand – e27.co/echelon
- Startup Thailand – startupthailand.org
- Techsauce Global Summit – summit.techsauce.co
- Thailand Digital Big Bang – digitalthailandbigbang.com

**Traditional Arts and Crafts**

- Baan Kang Wat – facebook.com/Baankangwat
- Baan Noorg Collaborative Arts and Culture – baannoorg.org
- The Jim Thompson Art Center – jimthompsonhouse.com

### Video Calls, Social Media, and Messaging Apps

- Facebook – facebook.com
- FaceTime – facetime.apple.com
- Google Duo – duo.google.com
- LINE – line.me/en/
- Skype – skype.com
- WhatsApp – whatsapp.com
- Zoom – zoom.us

### Volunteering

- BEES Elephant Sanctuary – bees-elesanctuary.org
- Care for Dogs – carefordogs.org
- Elephant Nature Park – elephantnaturepark.org
- Friends for Asia – volunteerthailand.org
- Gibbon Rehabilitation Project – gibbonproject.org
- GVI – gvi.co.uk
- Operation Smile Thailand – operationsmile.or.th
- Phuket Cleanse – phuketcleanse.com
- Phuket Sunshine Village Foundation – phuketsunshinevillage.org
- Projects Abroad – projects-abroad.org
- Scholars of Sustenance – scholarsofsustenance.org
- Soi Dog Foundation – soidog.org
- The Mercy Centre – mercycentre.org
- Trash Hero Thailand – trashhero.org
- Volunteers in Asia (VIA) – viaprograms.org

## Chapter 10: Family & Schooling

### Curriculum

- American Curriculum – ed.gov
- Australian Curriculum – australiancurriculum.edu.au

- British Curriculum (IGCSE and A-Levels) – cambridgeinternational.org
- International Baccalaureate (PYP) – ibo.org
- French Baccalaureate – aefe.gouv.fr
- German Abitur – kmk.org

**Educational Resources**

- Anki – apps.antiweb.net
- ChatGPT – openai.com
- Evernote – evernote.com
- Google Drive – google.com/drive
- MindMeister – mindmeister.com
- Slack – slack.com
- Trello – trello.com

**Expat Homeschooling Organisations**

- Families in Global Transition (FIGT) – figt.org
- Global Homeschool Exchange (GHEX) – ghex.world
- Worldschoolers – weareworldschoolers.org

**Language Support Apps**

- Babbel – babbel.com
- BBC Languages – bbc.co.uk/languages
- Duolingo – duolingo.com
- FluentU – flentu.com
- HelloTalk – hellotalk.com
- Italki – italki.com
- Memrise – memrise.com
- Mondly – mondly.com
- Pimsleur – pimsleur.com
- Rosetta Stone – rosettastone.com
- Tandem – tandem.net

**Local Meetups and Communities**

- Facebook Groups – facebook.com/groups
- InterNations – internations.org
- Meetup – meetup.com

**Online Study Platforms**

- Coursera – coursera.org
- EdX – edx.org
- Khan Academy – khanacademy.org
- Outschool – outschool.com
- Time4Learning – time4learning.com

**Schools**

- Bangkok Patana School – patana.ac.th
- British International School, Phuket (BISP) – bisphuket.ac.th
- Chiang Mai International School (CMIS) – cmis.ac.th
- Harrow International School Bangkok – harrowschool.ac.th
- HeadStart International School – headstartphuket.com
- Hua Hin International School – huahinschool.com
- International School Bangkok (ISB) – isb.ac.th
- International School Eastern Seaboard (ISE) – ise.ac.th
- Krabi International School – kis.com
- Lanna International School Thailand – lannaist.ac.th
- NIST International School – nist.ac.th
- Oonrak International School – oonraksamui.ac.th
- Panyaden International School – panyaden.ac.th
- Prem Tinsulanonda International School – ptis.ac.th
- Regents International School Pattaya – nordangliaeducation.com
- Ruamrudee International School (RIS) – rism.ac.th
- The International School of Asia – isa.ac.th
- UWC Thailand International School – uwcthailand.ac.th

# Author Bio

Hailing from the idyllic island of Tasmania (Australia), Candice Rafferty brings over 20 years of experience in Asia, including residencies in Bangkok, Beijing, Shanghai, Singapore, and Ho Chi Minh. Fluent in Chinese, she has a diverse background in media, hospitality, and public relations. Candice is an entrepreneur who has successfully founded two businesses and travelled to over 55 countries.

She holds an Associate Degree in Modern Languages (Mandarin), a Bachelor's degree in Classics & Archaeology and Chinese Studies, and a Master's degree in Communications. Additionally,-she is a qualified hatha and ashtanga yoga teacher.

Candice is known for her creativity, motivation, and team spirit, excelling in digital marketing, event planning, and storytelling. When she's not working, you can find her exploring new cultures, enjoying a good glass of wine, or perfecting her downward dog.

For more about her book and other forthcoming digital nomad titles, check out **@cetwixt.ink** on Instagram or send a message to **cetwixt.ink@gmail.com**